SpringerBriefs in Computer Science

For further volumes:
http://www.springer.com/series/10028

Abbas Jamalipour • Yaozhou Ma

Intermittently Connected Mobile Ad Hoc Networks

from routing to content distribution

foreword by H. Vincent Poor

 Springer

Abbas Jamalipour
The University of Sydney
School of Electrical and
Information Engineering
Sydney New South Wales
Australia
abbas.jamalipour@sydney.edu.au

Yaozhou Ma
The University of Sydney
School of Electrical and
Information Engineering
Sydney New South Wales
Australia
yaozhou.ma@sydney.edu.au

ISSN 2191-5768   e-ISSN 2191-5776
ISBN 978-1-4614-1571-8   e-ISBN 978-1-4614-1572-5
DOI 10.1007/978-1-4614-1572-5
Springer New York Dordrecht Heidelberg London

Library of Congress Control Number: 2011939060

Printed on acid-free paper

Springer is part of Springer Science+Business Media (www.springer.com)

to our families for their endless support ...

Foreword

Wireless networking is one of the most advanced and rapidly advancing technologies of our time. The modern wireless era has produced an array of technologies of tremendous economic and social value and almost ubiquitous market penetration. Many of these technologies, such as mobile phones and WiFi networks, are based on so-called infrastructure networks, in which information is transferred wirelessly between an end-user's device and an access point to a backbone network having a hierarchical control structure to manage information flow in the network. A more recently emerging network structure is that of an ad hoc network, in which end-user terminals communicate directly with one another in a peer-to-peer fashion without the benefit of a control structure provided by network infrastructure. Such networks are often formed in an ad hoc fashion (hence their name) as communicating devices appear in somewhat random locations. This type of structure presents new challenges in supporting efficient information flow, as protocols typically must rely on information forwarding through multiple intermediate devices, each of which acts autonomously, to deliver messages from source to destination. These challenges become even greater when the devices are moving, as the network topology then becomes dynamic. Even further challenges arise due to the fact that continuous end-to-end connectivity cannot be guaranteed in such networks, opening up the issues of what types of information content can be practically transferred through such networks and how this can best be accomplished. These latter issues are the subject of this excellent monograph, which collects for the first time in book form the considerable recent research dedicated to this important emerging topic, much of it by the authors themselves. Applications involving such *intermittently connected mobile ad hoc networks*, or ICMANs, are certain to multiply in the coming years as the technical sophistication and geographical spread of end-user devices increase, and thus this work by two leading contributors to the field should be of considerable interest to the researchers and engineers looking to develop the next generations of wireless networking technologies.

Princeton, New Jersey, August 2011   *H. Vincent Poor*

Preface

Mobile Ad Hoc Networks (MANETs) have changed the classical centralized wireless network topology into a whole new domain with many potential applications. While military applications of the MANET have been understood for some time, research on civil applications of the MANETs have become one of the most important topics among telecom engineers just in the beginning of this century.

In general, MANETs are the networks that are formed dynamically by an autonomous set of mobile nodes connected through wireless links without relying on any pre-configured infrastructure or a central base station. These nodes dynamically create the network connectivity through a temporary network topology, allowing them to seamlessly communicate with one another in areas with no pre-existing communication infrastructure. Owing to the self-configuration nature of MANETs where connectivity between peer end-terminals can be established automatically without any pre-configured infrastructure, data transfer is carried out through the cooperation among intermediate nodes over multi-hop routing paths. However, one of the key assumptions for such a routing approach, namely the existence of an end-to-end routing path, becomes untenable in cases where the network experiences intermittent connectivity due to limited transmission range, sporadic node densities, power limitations, and so on. To administer information tracking and data delivery capability in fostering service improvement in the emergency response and the eHealth sector, as well as pervasive computing in rural areas, it is necessary to develop a supplemental data dissemination framework based on opportunistic delivery probability. In fact, if the delivery delay can be tolerated, then node mobility can be exploited to deliver selected messages to the destination. Obviously, these so-called store-carry-forward (SCF) techniques incur considerable signaling overhead to include message summaries and meeting probabilities, resulting in the degradation of network efficiency.

Based on the above concept, a new subset of MANETs, called Intermittently Connected Mobile Ad Hoc Networks (ICMANs) could be created. By considering the nature of intermittent connectivity in most real world mobile environments without any restrictions placed on users' behavior, ICMANs are eventually formed without any assumption about the existence of an end-to-end path between any pair of nodes who are wishing to communicate. It is different from the conventional MANETs, which have been implicitly viewed as a connected graph with established complete paths between every pair of nodes. For the conventional MANETs, mobility of nodes is considered as a challenge and needs to be handled properly to enable seamless communication between nodes. However, to overcome intermittent connectivity in the ICMANs context, mobility is recognized as a critical component for data communications between the nodes that may never be part of the same connected portion of the network. This comes at the cost of additional and considerable delay in data forwarding, since data are often stored and carried by the intermediate nodes waiting for the mobility to generate the next forwarding opportunity that can probably take the data packet close to the destination. Such incurred large delays primarily limit ICMANs to the applications, which must tolerate delays beyond traditional forwarding delays. ICMANs belong to the family of delay tolerant networks (DTNs). However, their unique characteristics (e.g., self-organizing, random mobility and ad hoc based connection) derived from MANETs distinguish ICMANs from other typical DTNs such as interplanetary network (IPN) with infrastructure-based architecture.

By allowing mobile nodes to connect and disconnect based on their behaviors and wills, ICMANs enable a number of novel applications to become possible in the field of MANETs. For example, there is a growing demand for efficient architectures for deploying opportunistic content distribution systems over ICMANs. This is because a large number of smart handheld devices with powerful functions enable mobile users to utilize low cost wireless connectivity such as Bluetooth and IEEE 802.11 for sharing and exchanging the multimedia contents anytime anywhere. Note that such phenomenal growth of content-rich services has promoted a new kind of networking where the content is delivered from its source (referred to as publisher) towards interested users (referred to as subscribers) rather than towards pre-specified destinations. Comparing to the extensive research activities relating to the routing and forwarding issues in MANETs and even DTNs, opportunistic content distribution based on ICMAN is just in its early stage and has not been widely addressed.

This book covers the results of research carried out by the authors on the topic of ICMAN. This book provides an in-depth discussion on the latest research efforts for opportunistic content distribution over ICMANs. The discussion begins by introducing ICMANs, DTNs, and the most notable forwarding and routing technologies (e.g., epidemic routing and its variations, cluster-based routings and super-node-based routings). Chapter 1 also briefly discusses the mobility impact on routing performance; an important factor used in enabling packet routing in the network. Chapter 2 explains various forms of opportunistic content distributions over IC-MANs. It gives details about some proposed optimizing solutions by considering the mobility characteristics of the nodes. The idea of cooperation is further employed to allow the encountering nodes to work together for carrying out the cooperative decision-making strategies when the network resources become constrained. Apart from designing forwarding strategies to deliver the content from its publisher to their interested subscribers, content search or content lookup is another fundamental problem that determines the architecture and performance of opportunistic content distribution in the ICMANs context. Therefore, Chapter 3 focuses on illustrating an opportunistic content search mechanism to allow mobile users to discover interested contents based on the related keywords over ICMANs.

While there are many books written on the topic of MANET, this book provides a fresh and unique text on one of its main variant networks: the ICMAN. This book presents the main research results that we have produced on this topic in a self-contained and streamlined format that could be useful for readers who are interested in further research in the field. We believe that the future of wireless technology will be based on less centralized and more decentralized and distributed network topologies compared to the commonly existing cellular topologies, and in that sense ICMANs could become a major player within the next generation mobile network and content delivery infrastructures. We hope that this book provides sufficient inspiration to the endless potentials of the wireless technology and results in new networks and topologies to be created in near future.

Sydney, Australia,   *Abbas Jamalipour*
August 2011   *Yaozhou Ma*

Acknowledgements

This research was supported under Australian Research Council's Discovery Projects funding scheme (project number DP0985595).

Abbas Jamalipour is the Chair Professor of Ubiquitous Mobile Networking at the School of Electrical and Information Engineering, The University of Sydney, Australia. Yaozhou Ma is a Research Fellow at the School of Electrical and Information Engineering, The University of Sydney, Australia.

Contents

Acronyms

AHP	Analytical Hierarchy Process
AODV	Ad hoc On-demand Distance Vector
AP	Access Point
BLER	Bus Line based Effective Routing
CAN	Content Addressable Network
CAR	Context-aware Adaptive Routing
CCCDF	Cooperative Cache-based Content Distribution Framework
CDN	Content Delivery Network
CFS	Cooperative File System
CWC	Current Window Counter
DHT	Distributed Hash Table
DSDV	Destination-Sequenced Distance-Vector
DSR	Dynamic Source Routing
DTN	Delay Tolerant Network
EPL	Effective Path Length
ExOR	Extreme Opportunistic Routing
FIFO	First-In-First-Out
FLDF	Fuzzy Logic based Delivery Framework
GeRaF	Geographic Random Forwarding
GPS	Global Positioning System
GRA	Grey Relational Analysis
ICMAN	Intermittently Connected Mobile Ad hoc Network
IEEE	Institute of Electrical and Electronic Engineers
IID	Independent and Identical Distribution
IP	Internet Protocol
IPN	InterPlanetary Network
ISM	Intelligent Search Mechanism
LAN	Local Area Network
LSF	Last-Seen-First
LSI	Latent Semantic Indexing
MAC	Medium Access Control

MANET	Mobile Ad hoc NETwork
MED	Minimal Expected Delay
MEED	Minimum Estimated Expected Delay
MGOR	Multirate Geographic Opportunistic Routing
MIMO	Multiple Input Multiple Output
MMF	Most-Mobile-First
MORE	Mac-independent Opportunistic Routing and Encoding
MPCR	Minimum Power Cooperative Routing
MSF	Most-Social-First
MV	Meetings and Visits
NP	Nondeterministic Polynomial
ODE	Ordinary Differential Equation
OLSR	Optimized Link State Routing
OR	Opportunistic Routing
OSPF	Open Shortest Path First
P2P	Peer-to-Peer
PDA	Personal Digital Assistant
PREP	PRioritized Epidemic Routing
PRoPHET	PRobabilistic routing Protocol using History of Encounters and Transitivity
QoS	Quality of Service
RFID	Radio Frequency IDentification
RI	Routing Index
RIP	Routing Information Protocol
RLC	Random Linear Coding
RP	Rendezvous Point
SCAR	Sensor Context-Aware Routing
SCF	Store-Carry-Forward
SEPR	Shortest Expected Path Routing
SWIM	Shared Wireless Info-station Model
TCP	Transmission Control Protocol
TTL	Time-To-Live
VANET	Vehicular Ad hoc NETwork
VoIP	Voice-over-Internet Protocol
VSM	Vector Space Model
WAN	Wide Area Network
WMN	Wireless Mesh Network
ZRP	Zone Routing Protocol

Chapter 1
Introduction to Intermittently Connected Mobile Ad Hoc Networks

1.1 Overview

In the past decade, the emergence of cheap, small, powerful and smart wireless devices such as mobile phones, laptops, personal digital assistant (PDAs), etc have resulted in an exponential growth of mobile wireless networks. Currently, most of the connections among wireless devices are achieved over fixed infrastructure-based wireless networks. To mention only a few examples, cellular networks have been widely deployed to set up connections among mobile phones, while popularity of wireless access points (APs) allows travelers to surf the Internet from cafes, railway stations, airports, and other public locations. Although infrastructure-based wireless networks offer an effective way for mobile devices to get network services, setting up the infrastructure takes time and incurs potentially high costs. Moreover, there are scenarios where user-required infrastructure is not available, cannot be deployed in time, or cannot be deployed at all. Examples range from battlefield communications, disaster relief to wildlife tracking and habitat monitoring sensor networks. Providing the required network services in such scenarios leads to a mobile ad hoc network.

1.1.1 Mobile Ad Hoc Networks

In general, mobile ad hoc networks (MANETs) are networks formed dynamically by an autonomous set of mobile nodes that are connected via wireless links without relying on any pre-configured infrastructure or centralized administration. In other words, these participating wireless mobile nodes can freely and dynamically self-organize into arbitrary and temporary network topologies, allowing themselves to seamlessly communicate with each other in areas with no preexisting communication infrastructure. In MANETs, each node communicates directly with any other node within its transmission range, while communication beyond this range is established by employing intermediate nodes to set up a path in a hop-by-hop manner.

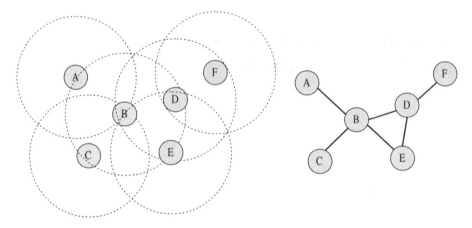

Fig. 1.1 Mobile ad hoc network

Figure. 1.1 shows an example of mobile ad hoc network and its communication topology. Note that due to the self-organized nature of MANETs, if these nodes are free to move randomly, they may organize themselves arbitrarily, and thereby the network topology can change rapidly and unpredictably. Therefore, Fig. 1.1 only represents a snapshot of network topology at a certain time instant.

In general, the characteristics of MANETs can be summarized as follows:

- Wireless;
- Mobility;
- Ad hoc based connection;
- Infrastructureless architecture;
- Multihop routing.

MANETs remove the constraints of infrastructure and allow devices to establish and join the network anywhere at any time. However, as mobile nodes are moving arbitrarily, the network topology may change constantly, resulting in route changes, frequent network partitions, and, even packet losses. Moreover, MANETs are normally formed by mobile devices with limited battery power, and the equipped wireless interfaces may also suffer limited bandwidth and high error rate. Hence, in order to accommodate such dynamic topology of MANETs as well as resource-constrained devices, an abundance of MANET routing protocols have been proposed in the literature. Based on the timing that the routes are established, these protocols can be classified into the following main categories:

- *Proactive routing* protocols: Each node propagates route updates proactively and periodically over the network to allow any other node to maintain a consistent and up-to-date routing table. Representative proactive protocols include optimized link state routing (OLSR) protocol [16] and destination-sequenced distance-vector (DSDV) protocol [70].

- *Reactive on – demand routing* protocols: A route to a destination, on the other hand, is established only when there is a demand from the source node. When needed, the source node triggers a path discovery process over the network to set up the route to the destination. Once the route has been set up properly, it is maintained either until it becomes no longer used or has expired, or until the destination becomes inaccessible from the source. Both dynamic source routing (DSR) [41] and ad hoc on-demand distance vector (AODV) routing [69] are referred to as representative examples of reactive on-demand routing protocols.
- *Hybrid routing* protocols: The characteristics of proactive routing protocols and reactive on-demand routing protocols are combined to form hybrid routing protocols. Zone routing protocol (ZRP) [32] is a typical example representing hybrid routing protocols.

1.1.2 Intermittently Connected Mobile Ad Hoc Networks

All the previously mentioned routing protocols implicitly assume that MANET is connected and there exists a complete end-to-end path between any pair of nodes wishing to communicate with each other. For example, as shown in Fig. 1.2, despite the network topology is dynamically changing over time, the existing routing protocols can always find the available end-to-end paths between nodes D and E and can subsequently facilitate data communication between them.

This assumption restricts those routing protocols to networks containing enough nodes to build a fully connected topology. Unfortunately, there exist some situations where intermittent connectivity may arise from node mobility, short transmission range, sporadic node density, power limitations, and so on, and thereby most of the time a complete end-to-end path between any two nodes does not exist, or such a path is highly unstable while being discovered. Fig. 1.3 is a typical example representing such situation. As can be seen, if node D still wants to deliver data to node E over time, the existing routing protocols for conventional MANETs cannot work or may suffer serious performance. This is because there is not any end-to-end path over time, which can be discovered by such routing protocols to facilitate the corresponding data communication.

A military mobile ad hoc network may become intermittently connected when mobile nodes (e.g., soldiers, tanks) move out of each other's transmission range and are subject to being destroyed. Moreover, pocket switched networks [12] are another similar examples, since they are formed by human carried mobile devices based on their Bluetooth or IEEE 802.11 interfaces with short transmission range. In addition, similar intermittent disconnectivity can be encountered by vehicular ad hoc networks (VANETs) due to high vehicle velocities (compared to the transmission range). In the literature, mobile ad hoc networks that suffer such intermittent connectivity (in the absence of end-to-end routing path) are commonly referred to as intermittently connected mobile ad hoc networks (ICMANs) [97].

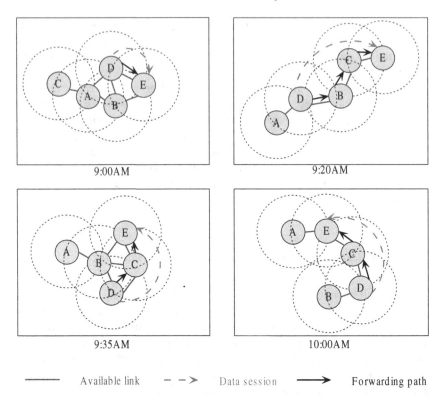

———	Available link	- -> Data session	——> Forwarding path

Fig. 1.2 An example of data communication in a typical mobile ad hoc network based on existing routing protocols

Real-time communications (e.g., voice-over-Internet Protocol (VoIP) and video streaming), which require a fully-connected path to forward sequenced packets timely, cannot work over ICMANs. However, it does not mean other applications can never be deployed over ICMANs. If delay can be tolerated by the applications, to improve the routing performance, i.e., to reduce packet loss ratio, mobility and storage spaces of intermediate nodes can be exploited to forward data to their final destinations. In other words, data can be temporarily stored in intermediate nodes until the node mobility generates the next possible forwarding opportunity. This is largely because connection topologies (constituting of mobile nodes) may overlap at different periods of time, thereby facilitating delayed data delivery to the destination over store-carry-forward strategy. Moreover, since any possible node can opportunistically be employed as the next carrier to bring data closer to its eventual destination, ICMANs are also referred to as opportunistic networks [67]. In addition, since incurred large delays primarily limit ICMANs to delay-tolerant applications, ICMANs belong to the family of delay tolerant networks (DTNs) [61].

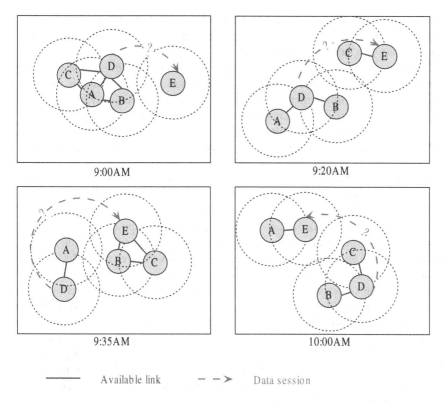

Fig. 1.3 An example of data communication in an intermittently connected mobile ad hoc network based on existing routing protocols

1.1.3 Delay Tolerant Networks

DTN has its root in interplanetary network (IPN) or interplanetary Internet [10][6] [22]. IPN was first designed to overcome some serious problems suffered by deep space communications between Earth and satellites, space stations, space probes and other spacecrafts where standard Internet protocols (i.e., TCP/IP protocols) are employed. This is because the TCP connections can easily be broken before any application data is transmitted [24] by the following problems:

- Narrow and highly asymmetric bandwidth, significant propagation delay and round trip time, and relatively high bit-error rate due to long distance between two communication nodes
- Frequent end-to-end disconnection because of planetary movement and occultation of communicating spacecrafts [61]

A typical IPN architecture across interplanetary space consists of an overlay network, called the bundle layer, above a wide range of regional networks. These regional networks are apart from each other and are connected through a number of gateways. To provide transparent communications for the application layer i.e., to hide disconnectivity and delays from the application layer, bundle layer consequently works in a store-forward manner [19].

There are some other types of networks, which may have a lot in common with IPN. An example of such networks is a village network, which is an intermittently connected infrastructure formed by low-cost wireless devices to provide rural connectivity for people in developing nations to access e-mail, voice mail, digital documents and other delay insensitive digital services [68][4]. For instance, in DakNet [68], a public bus equipped with a wireless interface can work as a mobile access point to provide intermittent connections between kiosks located in rural villages and Internet access point placed in the nearby town, while data is automatically exchanged when the bus is in transmission range of kiosks or Internet access point. DataMule [77] project for sensor networks is another example for such networks. In DataMule, sensors are spread over a large geographical area resulting a sparse sensor network, while a three-tier architecture is designed to provide connectivity between the sparse network and access points of wide area networks (WANs) by employing mobile entities such as people, animals or vehicles, which randomly move around the corresponding area. Similar approaches for sensor networks can be found in ZebraNet [44][57], shared wireless infostation model (SWIM) [31] and even featherlight information network with delay-endurable radio frequency identification (RFID) support (FINDERS) [93]. In addition, a sensor network may also suffer connection discontinuity, if sensors are scheduled to be wake/sleep periodically to conserve power. Based on the above observations, the IPN architecture is subsequently generalized to DTN architecture [6][22][11] to allow all these challenged networks to handle intermittent connectivity.

As can be seen, even though all these DTN applications primarily face the problem of intermittent connectivity and make use of store and forward based routing techniques, the unique characteristics derived from MANETs distinguish ICMANs from the other typical DTNs. For example, comparing to other infrastructure-based DTNs, ICMANs follow an infrastructureless-based architecture where every mobile node acts as a gateway mentioned earlier. In other words, every mobile node in IC-MANs is a router to make its own routing and forwarding decisions for all data it has stored whenever it suffers a transmitting opportunity.

1.2 Routing and Forwarding Techniques in ICMANs

Routing protocols such as RIP (Routing Information Protocol) and OSPF (Open Shortest Path First) used in IP networks, and AODV (Ad hoc On-demand Distance Vector), DSR (Dynamic Source Routing) and OLSR (Optimized Link State Routing) designed for the conventional MANETs cannot be deployed over ICMANs for data communication. This is because under these routing protocols, if the source node cannot discover the corresponding end-to-end path, the required data session cannot be facilitated (as shown in Fig. 1.3). Moreover, if the intermediate nodes cannot find a next hop for an arriving packet, the nodes also just simply discard the received packet, thereby disrupting the data transmission. As a result, to enable such nodes to exchange data between them and to allow some applications can employ ICMANs as the underlying networks, the main focus of research on ICMANs has been on routing and forwarding issues.

In ICMANs, connection topologies (constituting of mobile nodes) overlap at different periods of time and facilitate delayed data forwarding to the destinations over store-carry-forward (SCF) strategy. In other words, data is temporarily stored in the intermediate nodes until the node mobility generates the next possible forwarding opportunity. Accordingly, the general data forwarding in ICMANs consists of a sequence of local independent forwarding decisions made by the intermediate nodes. If the future network topology is known a priori or at least predictable, when and where to forward data can be determined ahead of time to achieve some optimal objective [39]. However, if the network topology is stochastic, mobility randomness and subsequently uncertainty of future state of the network should be addressed by SCF routing and forwarding strategies, even though node mobility is used to overcome the lack of end-to-end path.

By considering the network architectures formed over ICMANs, SCF routing and forwarding technologies can be classified as follows:

- *Super − node − based* approaches: In super-node-based approaches, additional powerful nodes with high storage capacity and energy are employed to participate in communication, thereby improving overall system performance. These super-nodes can be located at some specific geographical points, or move around in the network either randomly or following controlled trajectories.
- *Cluster − based* approaches: A number of ICMANs exhibits social network properties. Consequently, communities with strong intra-community connections can be formed virtually over ICMANs to improve forwarding efficiency in case of the future topology uncertainty.
- *Purely ad hoc* approaches: Rather than employing super-nodes or forming overlay clusters in the ICMANs context, majority of routing and forwarding approaches are designed in a completely flat ad hoc manner.

For purely ad hoc approaches, based on whether generating and forwarding copies of messages over the network to handle mobility uncertainty, they can further categorized as below:

- *Multi − copy or flooding − based* approaches: Due to the uncertainty of future network topology, majority of the SCF routing schemes send out more than one copy of each message to utilize more delivery opportunity and accordingly improve the successful delivery probability. For example, epidemic routing [87], proposed first for ICMANs, is a plain flooding based routing protocol where message copies diffuse over the network until at least one of them eventually reach the destination. Although optimal results can be achieved with unlimited resources, in most realistic cases with constrained resources such as buffer space, bandwidth and power, its performance is significantly degraded. A number of approaches have been introduced to reduce the cost and improve the performance of epidemic routing under the restrictions of network resources. In other words, the main objective of these approaches is to achieve message delivery rate with low delivery delay and overhead. These proposed approaches can be further categorized as below:

- *Estimation − based* approaches: In some realistic environments, especially human related scenarios (e.g., campus-life) where the nodes are likely to exhibit a repeated behavior pattern (not random), delivery overhead can be decreased by estimating the future delivery probability based on historical statistics to find better opportunities. Context describing the reality in which the user lives, and social relationships among users can also be utilized to predict the future connections. Note that the context could be information such as residence address, work address and addresses of sport and entertainment facilities.
- *Network − aware* approaches: In order to achieve the overall performance, apart from its individual delivery probability from mobility statistics and the context information, a node should also take into account the global knowledge of the network (e.g., the distribution of the current message copies existing over the network) when allocating the limited delivery opportunities.
- *Spraying routing* approaches. To significantly reduce the overhead of epidemic routing, while still maintaining good performance, these routing schemes allow the source of a message to put some limitations on the message to control the number of copies over the network.
- *Coding − based* approaches: Erasure coding and network coding from information theory can also be exploited to limit message flooding over the network.

- *Single − copy − based* approaches: Despite the increased robustness and low delay performance, multi-copy-based protocols consume a high amount of storage space, bandwidth, and energy. Consequently, it is still desirable for resource-constrained ICMANs to use single copy of messages to overcome the mobility uncertainty. Moreover, mobility information and even context information can also be utilized by single-copy-based approaches to select better next-hop to achieve the delivery rate and delivery delay.

A summary of the above classifications is shown in Fig. 1.4

1.2.1 Epidemic routing and its variations

To overcome the mobility uncertainty, the basic idea behind epidemic routing is to utilize every possible forwarding opportunity to deliver data to its eventual destination. For example, as shown in Fig. 1.5, a source, S, wishes to transmit a message to a destination D but no connected path is available at 9:00 AM. However, it can forward a copy of message to its encounter, C, and ask C to deliver the message copy in the future. Moreover, when S encounters B at 9:20 AM, it also asks B for help

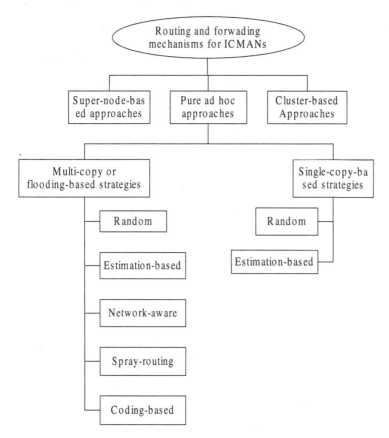

Fig. 1.4 Taxonomy of existing routing technologies for ICMANs.

by forwarding another message copy to *B*. At the same time, *C* is doing the same thing i.e., asking *E* to delivery another message copy in the future. Eventually, the message is delivered to its destination when *B* meets *D* at 9:35 AM, despite *S* and *D* cannot build a complete path between them during this period. As can be seen, epidemic routing relies on mobile nodes coming into contact through nodes mobility, since only when two mobile nodes are within their transmission range, they can exchange messages that the other node does not store.

As shown in Fig. 1.5, by replicating and sending as many message copies as possible through the connected portions of the network to find an optimal path, epidemic routing can maximize message delivery rate and minimize message delivery latency at the price of massive resource usage. To reduce the aggregated system resources consumed in message delivery, an upper bound on message hop count is placed to limit the number of epidemic exchanges that a particular message is subjected to. If the buffer size that each node is willing to allocate for epidemic routing is limited, the host will drop older messages in favor of newer ones upon reaching its capacity.

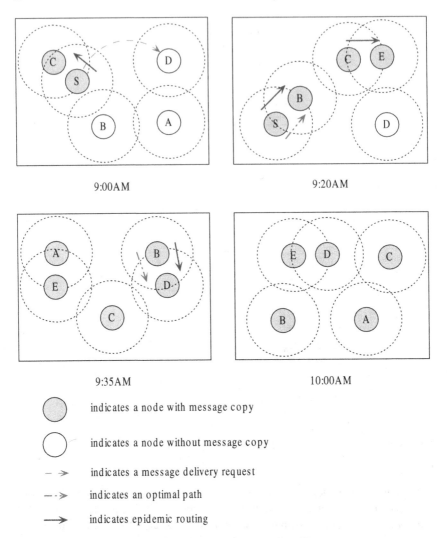

Fig. 1.5 Illustration of epidemic routing in an ICMAN.

Even optimal delivery performance in terms of delivery rate and delivery delay can be achieved by epidemic routing, it is resource hungry. Therefore, various queue policies and forwarding strategies are proposed for epidemic routing to get desirable performance in resource-constrained scenarios. Here queue policies aim to improve efficiency of storage utilization i.e. "Which messages should be dropped when the buffer is full?", while the goal of forwarding strategies is to determine "Which messages should be duplicated and forwarded when a node meets another node?" For example, "coin toss" strategy is introduced in [53] to determine a message should be forwarded or not. Rather than making random decisions, majority of the proposed approaches employ relative information to answer these two questions. Just as the first-in-first-out (FIFO) policy, some queueing and forwarding policies just needs knowledge about messages themselves. For instance, evict most forwarded first (MOFO) and evict shortest life time first (SHLI) queueing policies are introduced in [53] to drop the messages whenever the storage space becomes full. Note that MOFO policy requires the mobile node keep track of the number of copies each message has been forwarded, and the message with the largest number of forwarded copies is the first to be discarded, while in SHLI policy, the message with the shortest remaining lifetime is the first to be removed from the buffer. However, routing performance improves as more knowledge about the expected topology of the future and the existing message copies can be exploited. As a result, estimation-based approaches and network-aware approaches are illustrated in the following subsections.

1.2.1.1 Estimation-based semi-epidemic mechanisms

By exploiting the non-random mobility of the nodes existing in most human related environments, estimation-based routing mechanisms are the further step to answer the previously introduced questions. For example, in probabilistic routing protocol using history of encounters and transitivity (PRoPHET) [54], history of meeting statistics is collected by each intermediate node to calculate the delivery predictability for every other known node. The delivery predictability $P_{(a,b)}$ denoting the likelihood of a node a delivering a message to another node b is represented as follows:

$$P_{(a,b)} = \begin{cases} P_{(a,b)old} + (1 - P_{(a,b)old}) \times P_{init}, & a \text{ and } b \text{ encounter each other} \\ P_{(a,b)old} \times \gamma^k, & \text{otherwise} \end{cases} \quad (1.1)$$

where $P_{init} \in (0,1]$ is an initialization constant while $\gamma \in (0,1)$ is an aging constant, and k is the number of elapsed time units since the last time the metric was aged. A transitive property of the delivery predictability is also introduced in [54] by observing that if node a encounters b frequently while b encounters c frequently, a could be a good carrier to deliver messages destined for c. The influence of such transitivity on the delivery predictability can be represented as below,

$$P_{(a,c)} = P_{(a,c)old} + (1 - P_{(a,c)old}) \times P_{(a,b)} \times P_{(b,c)} \times \beta \quad (1.2)$$

where $\beta \in [0,1]$ is a scaling constant indicating the impact of the transitivity. Based on the delivery predictability, several estimation-based queueing policies such as evict most favorably forwarded first (MOPR) and evict least probable first (LEPR) and forwarding strategies including GRTR, GRTRSort and GRTRMax are illustrated in [53]. Note that in these forwarding strategies, when two nodes encounter each other, the message is only duplicated from a node with lower delivery predictability to the other with a higher delivery predictability.

Similar to the work in [54] [53], meetings and visits (MV) forwarding algorithm [7] employs information about meetings between mobile nodes and their visits to geographical locations to make forwarding decisions. In MV routing protocol , the probability that a node k can successfully deliver a message to region i with n transfers is calculated as follows:

$$P_n^k(i) = 1 - \prod_{j=1}^{N}(1 - m_{jk}P_{n-1}^j(i)) \tag{1.3}$$

where N is total number of nodes in the network, m_{jk} represents the probability of node j and k visiting the same region simultaneously, and $P_0^j(i)$ represents the probability of node j visiting the region i. Note that one assumption in MV routing protocol is that messages are delivered to stationary destinations located on a gird, therefore a message copy is only forwarded to an encounter with higher the probability of delivery. Additionally, MobySpace routing [48][49] employs mobility patterns of the nodes as the context information to make decisions. In MobySpace routing, a high dimensional Euclidean space is constructed upon mobility patterns of the nodes. Each axis in this space represents the frequency of being found in each possible location while the coordinate of a node indicates its mobility pattern. Hence, the similarity between mobility patterns of two nodes can be measured as a distance in the corresponding Euclidean space, and the best forwarding node for a message is the node that is as close as possible to the destination node in the space.

Shortest expected path routing (SEPR), another estimation-based approach proposed in [86], employs the link existing probability estimated based on history encountering statistics to carry out the shortest expected path length (E_{path}) at first. This is illustrated in Fig. 1.6 which depicts a stochastic model as a weighted complete graph. Note that associated link metrics (i.e., link forwarding probability) between any pair of nodes i and j is given by

$$P_{i,j} = \frac{Time_{connection}}{Time\_Winow} \tag{1.4}$$

where $Time\_Winow$ is the sliding sampling time window, while $Time_{connection}$ is the total time that i and j stay connected during $Time\_Winow$. According to this expression, Dijkstra algorithm can be performed to find the shortest E_{path} between any two nodes where it minimizes $\sum \frac{1}{P_{i,j}}$ i.e.,

$$E_{path} = \min \sum \frac{1}{P_{i,j}} \tag{1.5}$$

However, instead of simply using the shortest E_{path} to make forwarding decision, each message stored in the cache is then assigned an effective path length (EPL), which is set to infinite when it is a new received message. Accordingly when node A encounters B, its stored message m with the destination of node D is duplicated to B, only if

$$E_{path}(B,D) < \min(E_{path}(A,D), EPL_m(A)) \tag{1.6}$$

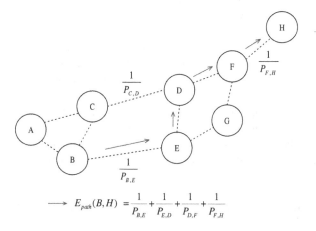

$$E_{path}(B,H) = \frac{1}{P_{B,E}} + \frac{1}{P_{E,D}} + \frac{1}{P_{D,F}} + \frac{1}{P_{F,H}}$$

Fig. 1.6 An example of stochastic network model.

Once the duplication is forwarded, $EPL_m(A)$ is updated by $E_{path}(B,D)$. Moreover, the queueing policy is also carried out according to EPL and E_{path}. That is, when the buffer space is full, the message with the smallest but not infinite EPL is discarded first, while if it does not exist, the message with the largest E_{path} is dropped first. As can be seen, in SEPR, EPL tracks the highest E_{path} of a message that a node has found before it meets the current encounter and only when E_{path} of the current encounter is greater than its EPL, the node will forward a message copy to its current encounter. Similar forwarding strategy is employed by delegation forwarding [21], which allows a message carrier to forward the message copies only to the encounters with higher qualities than all current message holders.

A variation of Dijkstra algorithm is also used by MaxProp protocol [5] to estimate delivery likelihood for each packet. Such estimated delivery likelihoods are used to define the order in which packets are transmitted and deleted. In addition to this, several complementary mechanisms are also considered by MaxProp to increase the delivery rate and to lower the delay of delivered packets. For example, acknowledgements are flooded over the network to notify the successfully delivered packets, while packets that have not traversed far in the network are given higher forwarding priorities. Moreover, MaxProp also attempts to prevent each node from receiving the same packet copy twice during a certain period. Similar ranking process can also be found in prioritized epidemic routing (PREP) [72] where drop priorities based on shortest path cost to the destinations are assigned to the messages having a hop count value greater than or equal to a configured threshold, while transmission priority is based on shortest path cost to the destination and the time-to-expire of message.

1.2.1.2 Network-aware-based semi-epidemic mechanisms

Although the delivery probability derived from the historical information offers a good metric for non-random mobility conditions in most realistic scenarios, when it is considered alone, the routing schemes fail to address the distribution of the existing message copies over the network. If more copies of a message have been replicated after it was generated, it signifies more intermediate nodes participating in the delivery process and a higher probability of delivery success. Therefore, several proposed routing approaches employ various utility functions to evaluate each possible messages based on mobility information, message distributions and even other related information.

Network-aware-based optimal buffer management policies for delay tolerant networks were introduced in [45]. For example, by assuming the inter-meeting time of the mobility model is exponentially distributed or has an exponential tail, the optimal buffer management policy to maximize the future delivery rate is to discard message i having the smallest value of the following utility:

$$(1 - \frac{m_i(T_i)}{L-1})\lambda R_i \exp(-\lambda n_i(T_i)R_i) \tag{1.7}$$

where L indicates total number of nodes in the network, λ is meeting rate between two mobile nodes, T_i and R_i represent elapsed time and remaining time for live for message i, respectively. In addition, $m_i(T_i)$ describes number of nodes (excluding source) have seen message i during T_i while $n_i(T_i)$ illustrates number of copies of message i in the network after T_i. Note that this utility actually represents the marginal utility value of a copy of message i regarding the total delivery rate. Moreover, to minimizing the average delivery delay, another utility function i.e.,

$$\frac{1}{n_i(T_i)^2\lambda}(1 - \frac{m_i(T_i)}{L-1}) \tag{1.8}$$

representing the marginal utility value for a copy of message i with respect to the average delivery delay can be employed to drop messages when the buffer becomes full. Similar optimal joint scheduling and drop policies are also illustrated in [46]. While by considering DTN routing as a resource allocation problem, a suboptimal policy by considering the existing message copies is introduced by RAPID [3] to schedule messages under constrained bandwidth.

Fuzzy logic based delivery framework (FLDF) [59][58] also enables every intermediate node to employ current number of message copies and historical encountering information to evaluate and rank all possible messages so as to allocate its buffer space. In order to address the mobility uncertainty and deal with the imprecise information due to the intermittent connectivity, fuzzy logical, an artificial intelligence technique proposed to handle problems with uncertain terms, is employed by FLDF to carry out the store-and-carry preference degree of each possible message (as shown in Fig. 1.7). Similar idea is also investigated in [60], where grey relational analysis (GRA) is utilized to perform message evaluation.

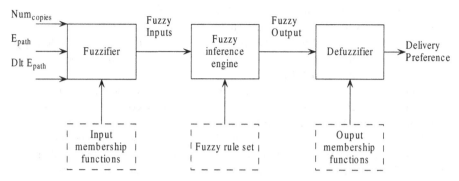

Fig. 1.7 Block diagram of the fuzzy logic system for FLDF.

1.2.1.3 Spray routing mechanisms

Alteratively, in order to reduce the overhead of epidemic routing, spray routing protocols proposed in [81][84][83] allow source node to give a carefully chosen total number of message copies to its generated message. Such carefully determined numbers ensures that the total number of transmissions is small and controlled.

For example, "Spray and Wait" [81][84] fist spreads L message copies to L distinct nodes, then if the destination is not in these L distinct nodes, all copies are stored and carried until one of these nodes encounters the destination. As can be seen, even though the spray phase of this mechanism spreads message copies in a manner similar to epidemic routing, it stops when the number of generated copies can guarantee that at least one copy will be directly forwarded to the destination. Note that such an idea has also been proved to be beneficial by [80]. Moreover, an optimal spraying strategy i.e., binary spraying, is introduced to minimize the expected time of distributing all L copies to their carriers when the movement of all mobile nodes follows an independent and identical distribution (IID). In binary spraying, any node holding more than one message copies will give half of the copies to a newly encountered node, even if it is not source. Consequently, the upper bound of the expected delay of "Spray and Wait" algorithm can eventually be represented as follows:

$$ED_{sw} \leq \sum_{i=1}^{L-1} \frac{EM_{mm}}{M-i} + \frac{M-L}{M-1} \frac{EM_{mm}}{L} \tag{1.9}$$

where M represents total number of nodes, L is the number of copies and EM_{mm} denotes the expected meeting time under the given mobility model. As a result, the source node can choose the number of copies (i.e., L) to achieve its required expected delay.

Even though "Spray and Wait" is simple and efficient, it requires enough nodes to roam around the whole network frequently. Therefore, if the mobility of each node is restricted to a local area, none of the carriers might ever encounter the destination. As a result, "Spray and Focus" [84] uses focus phase instead of wait phase. In focus phase, each copy is routed independently according to an utility-based single-copy scheme with transitivity [85]. The detail of such utility-based single-copy scheme will be given in subsection 1.2.2.

As mentioned earlier, binary spraying is optimal in a homogeneous environment. However, for a heterogenous environment, these L message copies need to be spread to L "better" carriers. Hence, several heuristic spraying strategies are proposed in [83]. In these strategies, any node with more than one message copies only give half of the copies to a newly encounter, which is considered to be "better" than the node itself based on some judgements. According to the judgements, these strategies are classified as last-seen-first (LSF) spraying, most-mobile-first (MMF) spraying and most-social-first (MSF) spraying. Moreover, encounter-based routing [65] introduces another intelligent spraying strategy, where the number of replicas of a message sent from one node to the other is based on their encounter values (EVs). For example, for two encountering nodes A and B, for a message M, node A sends

$$m \cdot \frac{EV_B}{EV_A + EV_B} \qquad (1.10)$$

copies of M to node B, where m is the total number of copies of message M while EV for each node is calculated as follows:

$$EV = \alpha \cdot \text{CWC} + (1 - \alpha) \cdot EV \qquad (1.11)$$

Note that current window counter (CWC) represents the number of encounters in the current time interval, while α is an exponential moving average weight.

1.2.1.4 Coding-based semi-epidemic mechanisms

Because of lack of knowledge of future topology, it is difficult to choose a single path (as mentioned earlier, consider topology overlapped over time) to deliver a message to its destination. As a result, to mitigate the effects of a single path failure, more than one identical copies of a message are sent by multi-copy-based approaches simultaneously over multiple paths. In other words, message redundancy is considered to cope with mobility uncertainty. However, as explained earlier, in most introduced approaches, message redundancy is just facilitated by simply generating several full copies of the message over the network. Therefore, in order to address such redundancy more efficiently, erasure coding techniques are accordingly adopted by researchers [89][40][51][15][88]. Note that in erasure coding, a message of k blocks is transformed into n ($n > k$) blocks such that the original message can be recovered from a subset of the n blocks, where $r = n/k$ is called as the replication factor and represents the level of redundancy.

The basic idea of using erasure coding over ICMANs can be found in [89], where the source of a message with k blocks first employs erasure code to generate the corresponding n code blocks and then equally splits such n code blocks among the first n encounters. After that, these block carriers are only allowed to send these blocks to the destination. The corresponding message can be decoded once any k blocks are collected by the destination. Actually, when $k = 1$, [89] is just same as other simple-replication approaches, since $r = n$ identical copies will be generated while only one copy is enough for the destination. The results in [89] shows that this very basic erasure coding based routing mechanism can provide the best worst-case delay performance with a fixed amount of overhead.

To cope with path failures and topology uncertainties, an optimization problem about how to allocate erasure code blocks over multiple paths to maximize the probability of successful message delivery is investigated in [40]. Two different path failure scenarios are considered: one is Bernoulli (0-1) path failure, which is formulated as a mixed integer problem and is proved to be nondeterministic polynominal (NP) hard, and the other is partial path failure, which is solved by using the modern portfolio theory. To further improve the performance of erasure-coding-based routing mechanisms, "Spray and Focus" like strategy is utilized by [51] to deliver the generated blocks to their destination. For spray phase, the portion of message blocks holden by the new encountering node should be proportional to its average contact frequency (ACF) to the destination, while in focus phase, the forwarding decision is also made by the value of ACF. Other erasure-coding-based routing approaches for DTNs and opportunistic networks can be found in [15][88].

Apart from erasure coding, network coding, another coding technique from information theory to achieve maximum throughput in a network, is also investigated. Comparing to erasure coding, where the source node splits a message into several blocks and distribute them over the network to control overhead, network coding allows an intermediate node to combine some of the received messages and send them out as a message to improve the network throughput. For example, for efficient communication in extreme networks, instead of simply forwarding the received messages, random linear coding (RLC) is employed by [91] to enable nodes to send out information coded over the contents of several messages they received. In addition, RLC is also utilized by [95][52] to improve the performance of epidemic routing, while [1] uses RLC in a human-related scenario, where a small percentage of nodes are connected to most of the other nodes. Moreover, the effect of erasure codes and fountain codes on the performance of the network is studied and analyzed in [2].

1.2.2 Single-copy-based routing mechanisms

Despite multi-copy-based routing approaches (the majority of routing schemes proposed for ICMANs with mobility uncertainty) can achieve delivery performance and robustness by taking advantage of node diversity, these protocols consume a high amount of resources such as storage space, power and bandwidth. Therefore, even though single-copy-based routing mechanisms introduce extra delivery delay

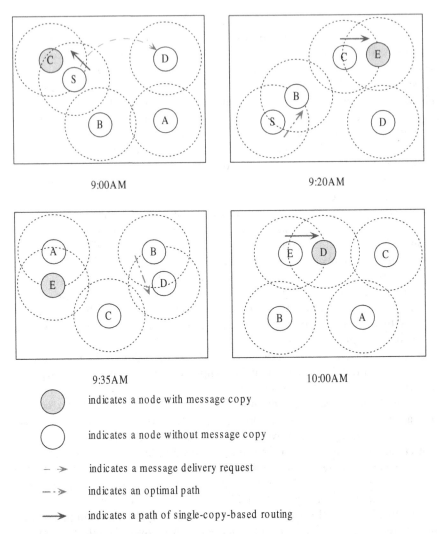

Fig. 1.8 Illustration of a single-copy-based routing in ICMAN.

because of difficulty of finding an optimal end-to-end path without the knowledge of future network topology (e.g., Fig. 1.8 vs Fig. 1.5), single-copy-based routing mechanisms still attract the attention of researchers for many resource-constrained ICMANs. Additionally, as mentioned earlier, in some spray routing algorithms (e.g., "Spray and Focus"), after a number of copies of a message are generated, each of them is forwarded independently based on a single-copy-based routing approach. For this reason, in order to design efficient multi-copy-based routing schemes, single-copy-based approaches should also be well investigated.

The simplest single-copy-based routing scheme is one-hop single-copy routing (i.e., direct transmission) [85], where a node forwards a message to another node it encounters, only if this encounter is the destination of this message. Even though this scheme is trivial, its expected delivery delay is an upper bound on the expected delay of any routing scheme for ICMANs and DTNs. For multi-hop single-copy-based routing schemes, a randomized algorithm is introduced in [85], where the current message holder hands over the message to another node it encounters with probability $p \in (0, 1]$. Reference [85] also shows that this simple routing strategy can result in expected progress for a number of mobility models such as Random Walk and Random Waypoint.

However, as illustrated earlier for multi-copy-based routing approaches, in order to improve delivery probability and to lower delivery delay, mobility information and even context information can also be considered by single-copy-based approaches to improve the decision-making accuracy for the next-hop selection. For example, to achieve the minimum estimated expected delay (MEED), a Dijkstra-based shortest path routing protocol using the observed contact history is proposed in [43][42]. The metric value of each edge or link in a topology graph is given as follows:

$$\frac{\sum_{i=1}^{n} d_i^2}{2t} \tag{1.12}$$

where n is the total number of disconnected periods, d_i is the duration of a given disconnected period, and t is the total time interval over which these disconnections were tracked. When local link-state information changes, the epidemic link-state protocol is used to distribute the latest information to all other nodes for routing table recalculation. MEED is an extension of minimal expected delay (MED), which is proposed in [39] for a scenario where the future network topology is known a priori.

Context-aware adaptive routing (CAR) [63][62] is another single-copy-based routing algorithm for delay-tolerant unicast communication in ICMANs. For a message, CAR uses a Kalman filter prediction and multi-criteria decision theory to choose the best next carrier. For a potential next carrier, the decision-making is carried out by an utility-function based on the parameters such as the change degree of connectivity of the node, its past contact with the destination of the message and its context information (e.g., battery level, memory availability or group membership). If information updates cannot be received on time, the future values of the context attributes and the delivery probabilities are predicted using a Kalman filter forecasting technique, originally developed in automatic control systems. The concept of CAR is also adopted by sensor context-aware routing (SCAR) [66][1]. Moreover, based on the time elapsed since any two nodes last saw each other, several utility-based single-copy forwarding strategies such as utility-based routing and utility-based routing with transitivity are introduced in [85] to choose the next hop. Additionally, a single-copy and multi-hop opportunistic routing mechanism aiming at minimizing delivery time in case of independent exponential pairwise inter-contacts can be found in [17].

---

[1] Note that SCAR is a multi-copy-based approach.

1.2.3 Cluster-based routing mechanisms

A number of ICMANs are formed by people, while the social relationships of people intend to be much more stable over time comparing to the network topology. Such social relationship implies that nodes in human-related ICMANs belong to different communities and the nodes among the same community tend to visit each other with higher probability compared to with other nodes outside this community. Recent collected human mobility traces from the real world have also proved this property [36][13][14]. Hence, forming virtual structured communities or clusters over ICMANs for efficient routing have attained the researchers' interests.

Reference [35] introduces a simple LABEL scheme, where each node is assumed to have a label that informs other nodes of its affiliation (i.e., the clusters and communities are predefined), while only the nodes belong to the same affiliation (i.e., same label) as the destination are selected as the message carriers. Similar idea is employed by [75] to design bus line based effective routing (BLER) for large-scale buses ad hoc networks. In BLER, each bus belongs to a bus line, while a message is first routed to the destination bus line based on Dijkstra algorithm, then the message is routed to the destination bus through the zigzag process i.e., the message is only forwarded to the bus running on the same route, but in the opposite direction.

Comparing to LABEL, where the community structure is defined in advance. A more comprehensive approach named as BUBBLE is introduced in [38]. BUBBLE considers two following assumptions:

- Every node belongs to at least one community, and
- every node has a global centrality (people, who interacts with more people, has higher centrality) across the whole network, and also a local centrality for each local community (a node may belong to multiple communities).

Based on such assumptions, routing mechanism of BUBBLE is carried out as follows: If a node has a message destined for another node, the message first bubbles up the hierarchical centrality tree based on the global ranking, until it reaches a node which belongs to the same community as the destination. Then the message continues to bubble up through the local centrality tree until the destination is reached or the message expires. Instead of using predefined communities or clusters, distributed community detection technologies including SIMPLE, K-CLIQUE and MODULARITY proposed in [37] are utilized by BUBBLE. Additionally, BUBBLE uses average unit-time degree, which counts unique nodes seen by a node during a unit-time, to approximate individual centrality value for each node. Centrality and similarity of nodes in a social network is also used by SimBet routing [18], a single-copy-based approach, for routing in ICMANs. Moreover, the distributions of node centrality investigated by [37][18][94] also prove that in the most social networks, only few people are more popular, and have connections with more people than the majority of the nodes.

LocalCom [50] also adopts the concept of structured community to design an epidemic forwarding scheme for DTNs. In LocalCom, while intra-community packet forwarding is done through single hop source routing and inter-community commu-

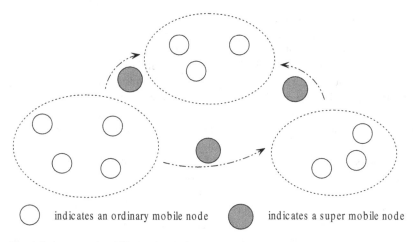

Fig. 1.9 An example of ICMAN with mobile super nodes.

nication employs controlled flooding. LocalCom uses virtual link based extended cliques to cluster mobile nodes into different communities. Such created communities have several desirable properties such as controllable diameter and strong intra-community connection. Moreover, to address routing scalability in a ICMAN with cyclic mobility, a DTN hierarchical routing (DHR) based on multilevel clustering can be found in [55][56].

1.2.4 Super-node-based routing mechanisms

For some applications such as data collection in sensor networks and Internet access in rural area, in order to help increasing possible connectivity and guaranteeing that isolated nodes can be reached, super-nodes, which are more powerful with respect to the ordinary nodes, are employed. These super nodes could be stationary at some specific geographical points or move around in the network area following either predetermined trajectories or arbitrary routes. For example, as shown in Fig. 1.9, mobile super nodes can be employed to offer communications among different network partitions.

In the message ferrying approach [98], additional mobile nodes named as message ferries are opportunistically deployed in a sparse MANET to provide a message relay service for message delivery. Two different message ferrying schemes are introduced: The first one is a node-initiated message ferrying, where the ferry moves based on a specific trajectory while each node in the network knows the ferry route and takes proactive movement to meet with the ferry when it has data to delivery. The second one is a ferry-initiated message ferrying, where the ferry take proactive movement to meet with the node after having received a service request from the node via a long range radio. Moreover, multiple ferries are used in [99] for communications among stationary regular nodes, which are disconnected from each other. This is largely because deploying multiple ferries can increase system throughput, reducing message delay and offer robustness to ferry failures. [99] focuses on designing ferry routes to meet the traffic demand with minimized data delivery delay, and presents ferry route algorithms for single ferry and multiple ferries cases, respectively. In the single ferry case, some algorithms for the well-known traveling salesman problem are exploited. For the multiple ferries case, other algorithms to assign nodes to specific ferries, synchronize among ferries, and assign ferries to specific routes are illustrated by considering different assumptions about the network. Such assumptions for the route design include whether the multiple ferries go over the same or different routes, and how ferries interact with each other.

Similarly, design of route and rendezvous point for message ferry and movement scheduling of both message and ferry are considered by logarithmic SCF routing [92]. Moreover, MORA routing [8] also introduces autonomous agents as additional participants into the network and these agents can adapt their movements in response to variations in network demand and capacity. Additionally, inter-regional messenger scheduling is addressed by [33] for DTNs.

Apart from deploying super mobile nodes as message ferries and agents, the problem of stationary relay nodes placement is addressed by [23] for vehicular delay-tolerant networks. Reference [23] proves that the problem of stationary relay nodes placement is an NP-hard problem and introduces heuristic algorithms to solve it. These algorithms have different objectives: MRH tries to minimize the number of relay nodes and the number of hop count while MRD aims to minimize the number of relay nodes and the average message delivery time.

Super-node-based routing approaches are widely considered by sensor networks for data collection as well. For example, DataMule system [77] exploits mobile entities to pick up data from the sensors, store it, and forward it to access points. It consists of a three-tier architecture:

- The bottom tier is made of the stationary wireless sensor nodes that periodically perform data sampling from the deployed area;
- The middle tier is composed of mobile agents named as MULEs, which have relatively large storage capacities and renewable power and move around in the sensor field to gather data and deliver it to the top tier;
- The top tier consists of access points that receive data from MULEs and deliver it to the WAN.

Apart from data collection in sensor networks, super-node-based routing mechanisms are also exploited to provide Internet access in rural area and developing regions (e.g., [68][76][29]), where kiosks in rural villages are intermittently connected to Internet access point in the nearby town by vehicles equipped with mobile access point.

1.3 Mobility Issues in ICMANs

As previously mentioned, mobility is a critical factor for the performance of IC-MANs. As a result, in this section, we first discuss the mobility impact on the SCF routing performance, followed by the latest efforts on the emerging human-related mobility models.

1.3.1 Mobility Impact on Routing Performance

Recently, [28] has proved that mobility can be employed by a two-hop relay algorithm to increase the capacity of a well connected MANET beyond the well-known theoretical limit found in [30], if extra delay can be tolerated. Based on this observation, a significant effort has already been devoted to understand the capacity-delay relationship in MANETs by addressing different mobility models. These mobility models include IID mobility model [64], Random Walk mobility model [25], Random Direction mobility model [79], Random Waypoint mobility model and Brownian mobility model [78][2]. At the same time, in order to overcome the underlying intermittent connectivity of ICMANs, node mobility is exploited to carry messages around the network as part of the message delivery. Therefore, the issue regarding how the mobility can affect the routing performance has also attained a lot of interest.

As message forwarding and transmitting in the ICMANs context occur only when mobile nodes encounter each other and the message transmission itself is significantly faster than nodes movement, the time elapsed between these meetings dominates end-to-end delay. Therefore, in order to investigate impact of mobility on the routing mechanisms, statistic characteristics of involved mobility models need to be studied first. It has been shown the meeting time in many popular mobility models can be modeled as a Poisson process i.e., the meeting time is exponentially distributed with mean $1/\lambda$ or has a least an exponential tail [45]. For example, the meeting time in Random Direction and Random Waypoint mobility models have been proved to follow the exponential distribution by [27][82]. Note that for Random Direction and Random Waypoint mobility models, [27] gives the estimations of λ which can be represented as:

$$\lambda_{RD} \approx \frac{2rE[V^*]}{L^2},\tag{1.13}$$

$$\lambda_{RW} \approx \frac{2\omega rE[V^*]}{L^2},\tag{1.14}$$

respectively, for values of $r \ll L$. Here r is transmission range, L^2 represents the area of a limited region where nodes move in, $\omega \approx 1.3683$ is a constant for the Random

---

[2] An exhaustive survey of mobility models can be found in Reference [9].

Waypoint model and $E[V^*]$ indicates the expected relative speed between two nodes. Similar derivation of the expected meeting time for such two mobility models can be found in [82]. Moreover, [82] also shows that the meeting time in community-based mobility model is exponentially distributed as well. Note that the community-based mobility model is represented as a simple two-state Markov Chain, where Random Waypoint mobility or Random Direction mobility can be performed by each node during each of these states. Apart from these "epoch-based" mobility models like Random Waypoint, Random Direction, and community-based mobility, [85] proves that the meeting time of Random Walk mobility model for a network area with large size has an exponential tail.

Based on such observations of mobility models, [27] exploits a Markov chain to carry out the expected message delay for two-hop multi-copy-based protocol and unrestricted multi-copy-based protocol (i.e., epidemic routing), which are given by

$$E[T_2] = \frac{1}{\lambda}(\sqrt{\frac{\pi}{2N}} + O(\frac{1}{N})), \qquad (1.15)$$

$$E[T_U] = \frac{1}{\lambda N}(\log(N) + \gamma + O(\frac{1}{N})), \qquad (1.16)$$

respectively, where N is the number of nodes excluding the destination node and γ is the Euler constant. Apart from the expected delivery delay, the expected number of message copies for such two routing schemes is also derived in [27] as follows:

$$E[N_2] = \sqrt{\frac{\pi N}{2}} + O(1), \qquad (1.17)$$

$$E[N_U] = \frac{N+1}{2}, \qquad (1.18)$$

Instead of using the Markov chain, [96] exploits ordinary differential equations (ODEs) to study epidemic routing and its variations and finds similar results.

1.3.2 Human-related Mobility Models

In terms of ICMAN, there are a lot of scenarios are human-related, and thereby exhibiting social characteristics, which are difficult to be justified by conventional random mobility models. Therefore, realistic traces collected from actual networks with mobile devices are considered by the researchers to evaluate the routing protocols and applications. For example, [45] uses the traces from ZebraNet project [90] and from taxi cabs in San Francisco city [71] to evaluate the proposed optimal buffer management policies, while the performance of MaxProp [5] and RAPID [3] are evaluated over UMass DieselNet. Additionally, SimBet Routing [18] is simulated by using traces from the MIT Reality Mining project [20].

However, it is difficult to deploy a real mobile network with large scale for each researcher to carry out its meaningful performance evaluation. Therefore, the collected mobility traces have been further analyzed, characterized and modeled for proposing new synthetic mobility models for simulation. Note that such investigation is also helpful for the researchers to better understand the fundamental properties of ICMANs. Based on the data sets collected from the Haggle project [74], [13][14] observe an approximately power-law distribution of inter-contact time (the time gap separating two successive contracts between the same pair of nodes). Similarly, by using mobility tracks collected from university campuses, theme park, and even a metropolitan area by using GPS devices, [73] discovered that the human mobility is close to Levy walk, where the flight and pause time distribution follow the truncated power-law. Accordingly, the authors propose a Levy-walk-based mobility model in the same paper. Apart from the Levy walk, some other important features such as heterogeneously bounded area and fractal waypoints have also been highlighted for human mobility [47]. Note that heterogeneously bounded area means that different people may have different area preferences and they have higher probabilities to visit their preferred areas. Such a property also can be found in [26] and [34]. Comparing to heterogeneously bounded area, the property of fractal waypoints implies that there always exist some popular places which can attract the majority of the people. By addressing all such features, a novel synthetic mobility model named as Self-similar Least-Action Walk (SLAW) has been proposed in [47]. Time-variant community (TWC) mobility model [34] is another recent proposed mobility model addressing spatial and temporal correlations in the human mobility patterns. Similar to heterogeneously bounded area and fractal waypoints, location visiting preferences and periodical reappearance of each node are considered by TWC mobility model.

1.4 Summary and Conclusions

This Chapter ws started with an overview of ICMAN and its relationships with MANET and DTN. After that, an in-depth review of the current state-of-the-art literature published in the area of routing technologies was provided. Amongst the discussed routing schemes were epidemic routing and its variations, the single-copy-based approaches, cluster-based approaches and super-node-based approaches. Finally the Chapter was concluded by discussing the mobility impact on the routing performance and the emerging social mobility models by considering human behaviors.

References

[1] Ahmed S, Kanhere SS (2009) Hubcode: message forwarding using hub-based network coding in delay tolerant networks. In: MSWiM '09: Proceedings of the 12th ACM international conference on Modeling, analysis and simulation

of wireless and mobile systems, ACM, Tenerife, Canary Islands, Spain, pp 288–296, DOI http://doi.acm.org/10.1145/1641804.1641853

[2] Altman E, De Pellegrini F (2009) Forward correction and fountain codes in delay tolerant networks. In: INFOCOM 2009, IEEE, Rio de Janeiro, Brazil, pp 1899–1907, DOI 10.1109/INFCOM.2009.5062111

[3] Balasubramanian A, Levine B, Venkataramani A (2007) Dtn routing as a resource allocation problem. In: SIGCOMM '07: Proceedings of the 2007 conference on Applications, technologies, architectures, and protocols for computer communications, ACM, Kyoto, Japan, pp 373–384, DOI http://doi.acm.org/10.1145/1282380.1282422

[4] Brewer E, Demmer M, Du B, Ho M, Kam M, Nedevschi S, Pal J, Patra R, Surana S, Fall K (2005) The case for technology in developing regions. Computer, IEEE 38(6):25–38, DOI 10.1109/MC.2005.204

[5] Burgess J, Gallagher B, Jensen D, Levine BN (2006) Maxprop: Routing for vehicle-based disruption-tolerant networks. In: INFOCOM 2006. 25th IEEE International Conference on Computer Communications. Proceedings, Barceloba, Spain, pp 1–11, DOI 10.1109/INFOCOM.2006.228

[6] Burleigh S, Hooke A, Torgerson L, Fall K, Cerf V, Durst B, Scott K, Weiss H (2003) Delay-tolerant networking: an approach to interplanetary internet. Communications Magazine, IEEE 41(6):128–136, DOI 10.1109/MCOM.2003.1204759

[7] Burns B, Brock O, Levine B (2005) Mv routing and capacity building in disruption tolerant networks. In: INFOCOM 2005. 24th Annual Joint Conference of the IEEE Computer and Communications Societies. Proceedings IEEE, Miami, FL, USA, vol 1, pp 398–408 vol. 1, DOI 10.1109/INFCOM.2005.1497909

[8] Burns B, Brock O, Levine BN (2008) Mora routing and capacity building in disruption-tolerant networks. Ad Hoc Networks 6(4):600 – 620, DOI DOI: 10.1016/j.adhoc.2007.05.002

[9] Camp T, Boleng J, Davies V (2002) A survey of mobility models for ad hoc network research. Wireless Communications and Mobile Computing 2(5):483–502, DOI 10.1002/wcm.72

[10] Cerf V, Burleigh S, Hooke A, Torgerson L, Durst R, Scott K, Travis E, Weiss H (2001) Interplanetary internet (ipn): Architectural definition. Http://www.ipnsig.org/reports/memo-ipnrg-arch-00.pdf, accessed in December 2009

[11] Cerf V, Burleigh S, Hooke A, Torgerson L, Durst R, Scott K, Fall K, Weiss H (2007) Delay-tolerant networking architecture. RFC 4838, IETF Network Working Group, http://www.apps.ietf.org/rfc/rfc4838.html, accessed in December 2009

[12] Chaintreau A, Hui P, Crowcroft J, Diot C, Gass R, Scott J (2005) Pocket switched networks: Real-world mobility and its consequences for opportunistic forwarding. Tech. Rep. 617, Computer Laboratory, the University of Cambridge, http://www.cl.cam.ac.uk/TechReports/UCAM-CL-TR-617, accessed in December 2009

[13] Chaintreau A, Hui P, Crowcroft J, Diot C, Gass R, Scott J (2006) Impact of human mobility on the design of opportunistic forwarding algorithms. In: INFOCOM 2006. 25th IEEE International Conference on Computer Communications. Proceedings, Barcelona, Spain, pp 1–13, DOI 10.1109/INFOCOM.2006.172

[14] Chaintreau A, Hui P, Crowcroft J, Diot C, Gass R, Scott J (2007) Impact of human mobility on opportunistic forwarding algorithms. Mobile Computing, IEEE Transactions on 6(6):606–620, DOI 10.1109/TMC.2007.1060

[15] Chen LJ, Yu CH, Sun T, Chen YC, Chu Hh (2006) A hybrid routing approach for opportunistic networks. In: CHANTS '06: Proceedings of the 2006 SIGCOMM workshop on Challenged networks, ACM, Pisa, Italy, pp 213–220, DOI http://doi.acm.org/10.1145/1162654.1162658

[16] Clausen   T,   Jacquet   P   (2003)   Optimized   link   state   routing protocol   (olsr).   RFC   3626,   IETF   Network   Working   Group, http://www.apps.ietf.org/rfc/rfc3626.html, accessed in December 2009

[17] Conan V, Leguay J, Friedman T (2008) Fixed point opportunistic routing in delay tolerant networks. Selected Areas in Communications, IEEE Journal on 26(5):773–782, DOI 10.1109/JSAC.2008.080604

[18] Daly EM, Haahr M (2007) Social network analysis for routing in disconnected delay-tolerant manets. In: MobiHoc '07: Proceedings of the 8th ACM international symposium on Mobile ad hoc networking and computing, ACM, Montreal, Quebec, Canada, pp 32–40, DOI http://doi.acm.org/10.1145/1288107.1288113

[19] Daly EM, Haahr M (2010) The challenges of disconnected delay-tolerant manets. Ad Hoc Networks 8(2):241 – 250, DOI DOI: 10.1016/j.adhoc.2009.08.003

[20] Eagle N, Pentland AS (2005) CRAWDAD data set mit/reality (v. 2005-07-01). Downloaded from http://crawdad.cs.dartmouth.edu/mit/reality, accessed in December 2009

[21] Erramilli V, Crovella M, Chaintreau A, Diot C (2008) Delegation forwarding. In: MobiHoc '08: Proceedings of the 9th ACM international symposium on Mobile ad hoc networking and computing, ACM, Hong Kong, China, pp 251–260, DOI http://doi.acm.org/10.1145/1374618.1374653

[22] Fall K, Farrell S (2008) Dtn: an architectural retrospective. Selected Areas in Communications, IEEE Journal on 26(5):828–836, DOI 10.1109/JSAC.2008.080609

[23] Farahmand F, Cerutti I, Patel A, Zhang Q, Jue J (2008) Relay node placement in vehicular delay-tolerant networks. In: Global Telecommunications Conference, 2008. IEEE GLOBECOM 2008. IEEE, New Orleans, LA, USA, pp 1–5, DOI 10.1109/GLOCOM.2008.ECP.483

[24] Farrell S, Cahill V, Geraghty D, Humphreys I, McDonald P (2006) When tcp breaks: Delay- and disruption- tolerant networking. Internet Computing, IEEE 10(4):72–78, DOI 10.1109/MIC.2006.91

[25] Gamal A, Mammen J, Prabhakar B, Shah D (2004) Throughput-delay tradeoff in wireless networks. In: INFOCOM 2004. Twenty-third AnnualJoint Con-

ference of the IEEE Computer and Communications Societies, Hong Kong, China, vol 1, p 475, DOI 10.1109/INFCOM.2004.1354518

[26] Gonzalez MC, Hidalgo CA, Barabasi AL (2008) Understanding individual human mobility patterns. Nature 453(7196):779–782, DOI 10.1038/nature06958

[27] Groenevelt R, Nain P, Koole G (2005) The message delay in mobile ad hoc networks. Performance Evaluation 62(1-4):210 – 228, DOI DOI: 10.1016/j.peva.2005.07.018, performance 2005

[28] Grossglauser M, Tse D (2002) Mobility increases the capacity of ad hoc wireless networks. Networking, IEEE/ACM Transactions on 10(4):477 – 486, DOI 10.1109/TNET.2002.801403

[29] Guo S, Falaki MH, Oliver EA, Ur Rahman S, Seth A, Zaharia MA, Keshav S (2007) Very low-cost internet access using kiosknet. SIGCOMM Comput Commun Rev 37(5):95–100, DOI http://doi.acm.org/10.1145/1290168.1290181

[30] Gupta P, Kumar P (2000) The capacity of wireless networks. Information Theory, IEEE Transactions on 46(2):388 –404, DOI 10.1109/18.825799

[31] Haas Z, Small T (2006) A new networking model for biological applications of ad hoc sensor networks. Networking, IEEE/ACM Transactions on 14(1):27–40, DOI 10.1109/TNET.2005.863461

[32] Haas ZJ, Pearlman MR, Samar P (2002) The zone routing protocol (zrp) for ad hoc networks. Draft-ietf-manet-zone-zrp-04, http://tools.ietf.org/id/draft-ietf-manet-zone-zrp-04.txt, accessed in December 2009

[33] Harras K, Almeroth K (2006) Inter-regional messenger scheduling in delay tolerant mobile networks. In: World of Wireless, Mobile and Multimedia Networks, 2006. WoWMoM 2006. International Symposium on a, Niagara-Falls, Buffalo-NY, USA, pp 10 pp.–102, DOI 10.1109/WOWMOM.2006.53

[34] Hsu WJ, Spyropoulos T, Psounis K, Helmy A (2009) Modeling spatial and temporal dependencies of user mobility in wireless mobile networks. Networking, IEEE/ACM Transactions on 17(5):1564 –1577, DOI 10.1109/TNET.2008.2011128

[35] Hui P, Crowcroft J (2007) How small labels create big improvements. In: Pervasive Computing and Communications Workshops, 2007. PerCom Workshops '07. Fifth Annual IEEE International Conference on, White Plains, NY, USA, pp 65–70, DOI 10.1109/PERCOMW.2007.55

[36] Hui P, Chaintreau A, Scott J, Gass R, Crowcroft J, Diot C (2005) Pocket switched networks and human mobility in conference environments. In: WDTN '05: Proceedings of the 2005 ACM SIGCOMM workshop on Delay-tolerant networking, ACM, Philadelphia, Pennsylvania, USA, pp 244–251, DOI http://doi.acm.org/10.1145/1080139.1080142

[37] Hui P, Yoneki E, Chan SY, Crowcroft J (2007) Distributed community detection in delay tolerant networks. In: MobiArch '07: Proceedings of 2nd ACM/IEEE international workshop on Mobility in the evolving internet architecture, ACM, Kyoto, Japan, pp 1–8, DOI http://doi.acm.org/10.1145/1366919.1366929

[38] Hui P, Crowcroft J, Yoneki E (2008) Bubble rap: social-based forwarding in delay tolerant networks. In: MobiHoc '08: Proceedings of the 9th ACM international symposium on Mobile ad hoc networking and computing, ACM, Hong Kong, Hong Kong, China, pp 241–250, DOI http://doi.acm.org/10.1145/1374618.1374652

[39] Jain S, Fall K, Patra R (2004) Routing in a delay tolerant network. In: SIGCOMM '04: Proceedings of the 2004 conference on Applications, technologies, architectures, and protocols for computer communications, ACM, Portland, Oregon, USA, pp 145–158, DOI http://doi.acm.org/10.1145/1015467.1015484

[40] Jain S, Demmer M, Patra R, Fall K (2005) Using redundancy to cope with failures in a delay tolerant network. In: SIGCOMM '05: Proceedings of the 2005 conference on Applications, technologies, architectures, and protocols for computer communications, ACM, Philadelphia, Pennsylvania, USA, pp 109–120, DOI http://doi.acm.org/10.1145/1080091.1080106

[41] Johnson DB, Maltz DA, Broch J (2007) The dynamic source routing protocol (dsr) for mobile ad hoc networks for ipv4. RFC 4728, IETF Network Working Group, http://www.apps.ietf.org/rfc/rfc4728.html, accessed in December 2009

[42] Jones E, Li L, Schmidtke J, Ward P (2007) Practical routing in delay-tolerant networks. Mobile Computing, IEEE Transactions on 6(8):943–959, DOI 10.1109/TMC.2007.1016

[43] Jones EPC, Li L, Ward PAS (2005) Practical routing in delay-tolerant networks. In: WDTN '05: Proceedings of the 2005 ACM SIGCOMM workshop on Delay-tolerant networking, ACM, Philadelphia, Pennsylvania, USA, pp 237–243, DOI http://doi.acm.org/10.1145/1080139.1080141

[44] Juang P, Oki H, Wang Y, Martonosi M, Peh LS, Rubenstein D (2002) Energy-efficient computing for wildlife tracking: design tradeoffs and early experiences with zebranet. In: ASPLOS-X: Proceedings of the 10th international conference on Architectural support for programming languages and operating systems, ACM, San Jose, California, pp 96–107, DOI http://doi.acm.org/10.1145/605397.605408

[45] Krifa A, Baraka C, Spyropoulos T (2008) Optimal buffer management policies for delay tolerant networks. In: Sensor, Mesh and Ad Hoc Communications and Networks, 2008. SECON '08. 5th Annual IEEE Communications Society Conference on, San Francisco, CA, USA, pp 260–268, DOI 10.1109/SAHCN.2008.40

[46] Krifa A, Barakat C, Spyropoulos T (2008) An optimal joint scheduling and drop policy for delay tolerant networks. In: World of Wireless, Mobile and Multimedia Networks, 2008. WoWMoM 2008. 2008 International Symposium on a, Newport Beach, CA, USA, pp 1–6, DOI 10.1109/WOWMOM.2008.4594889

[47] Lee K, Hong S, Kim SJ, Rhee I, Chong S (2009) Slaw: A new mobility model for human walks. In: INFOCOM 2009, IEEE, pp 855 –863, DOI 10.1109/INFCOM.2009.5061995

[48] Leguay J, Friedman T, Conan V (2005) Dtn routing in a mobility pattern space. In: WDTN '05: Proceedings of the 2005 ACM SIGCOMM workshop on Delay-tolerant networking, ACM, Philadelphia, Pennsylvania, USA, pp 276–283, DOI http://doi.acm.org/10.1145/1080139.1080146

[49] Leguay J, Friedman T, Conan V (2006) Evaluating mobility pattern space routing for dtns. In: INFOCOM 2006. 25th IEEE International Conference on Computer Communications. Proceedings, Barcelona, Spain, pp 1–10, DOI 10.1109/INFOCOM.2006.299

[50] Li F, Wu J (2009) Localcom: A community-based epidemic forwarding scheme in disruption-tolerant networks. In: Sensor, Mesh and Ad Hoc Communications and Networks, 2009. SECON '09. 6th Annual IEEE Communications Society Conference on, Rome, Italy, pp 1–9, DOI 10.1109/SAHCN.2009.5168942

[51] Liao Y, Tan K, Zhang Z, Gao L (2006) Estimation based erasure-coding routing in delay tolerant networks. In: IWCMC '06: Proceedings of the 2006 international conference on Wireless communications and mobile computing, ACM, Vancouver, British Columbia, Canada, pp 557–562, DOI http://doi.acm.org/10.1145/1143549.1143660

[52] Lin Y, Liang B, Li B (2007) Performance modeling of network coding in epidemic routing. In: MobiOpp '07: Proceedings of the 1st international MobiSys workshop on Mobile opportunistic networking, ACM, San Juan, Puerto Rico, pp 67–74, DOI http://doi.acm.org/10.1145/1247694.1247709

[53] Lindgren A, Phanse K (2006) Evaluation of queueing policies and forwarding strategies for routing in intermittently connected networks. In: Communication System Software and Middleware, 2006. Comsware 2006. First International Conference on, Delhi, India, pp 1–10, DOI 10.1109/COMSWA.2006.1665196

[54] Lindgren A, Doria A, Schelén O (2004) Probabilistic routing in intermittently connected networks. In: Service Assurance with Partial and Intermittent Resources, Lecture Notes in Computer Science, vol 3126/2004, Springer Berlin / Heidelberg, pp 239–254, DOI 10.1007/b99076

[55] Liu C, Wu J (2007) Scalable routing in delay tolerant networks. In: MobiHoc '07: Proceedings of the 8th ACM international symposium on Mobile ad hoc networking and computing, ACM, Montreal, Quebec, Canada, pp 51–60, DOI http://doi.acm.org/10.1145/1288107.1288115

[56] Liu C, Wu J (2009) Scalable routing in cyclic mobile networks. Parallel and Distributed Systems, IEEE Transactions on 20(9):1325–1338, DOI 10.1109/TPDS.2008.218

[57] Liu T, Sadler CM, Zhang P, Martonosi M (2004) Implementing software on resource-constrained mobile sensors: experiences with impala and zebranet. In: MobiSys '04: Proceedings of the 2nd international conference on Mobile systems, applications, and services, ACM, Boston, MA, USA, pp 256–269, DOI http://doi.acm.org/10.1145/990064.990095

[58] Ma Y, Jamalipour A (2009) Optimized message delivery framework using fuzzy logic for intermittently connected mobile ad hoc networks.

Wireless Communications and Mobile Computing 9(4):501–512, DOI 10.1002/wcm.693

[59] Ma Y, Kibria M, Jamalipour A (2008) A fuzzy logic-based delivery framework for optimized routing in mobile ad hoc networks. In: Wireless Communications and Mobile Computing Conference, 2008. IWCMC '08. International, Crete Island, Greece, pp 801–806, DOI 10.1109/IWCMC.2008.138

[60] Ma Y, Rubaiyat Kibria M, Jamalipour A (2008) Optimized routing framework for intermittently connected mobile ad hoc networks. In: Communications, 2008. ICC '08. IEEE International Conference on, Beijing, China, pp 3171–3175, DOI 10.1109/ICC.2008.597

[61] McMahon A, Farrell S (2009) Delay- and disruption-tolerant networking. Internet Computing, IEEE 13(6):82–87, DOI 10.1109/MIC.2009.127

[62] Musolesi M, Mascolo C (2009) Car: Context-aware adaptive routing for delay-tolerant mobile networks. Mobile Computing, IEEE Transactions on 8(2):246–260, DOI 10.1109/TMC.2008.107

[63] Musolesi M, Hailes S, Mascolo C (2005) Adaptive routing for intermittently connected mobile ad hoc networks. In: World of Wireless Mobile and Multimedia Networks, 2005. WoWMoM 2005. Sixth IEEE International Symposium on a, Taormina - Giardini Naxos, Italy, pp 183–189, DOI 10.1109/WOWMOM.2005.17

[64] Neely M, Modiano E (2005) Capacity and delay tradeoffs for ad hoc mobile networks. Information Theory, IEEE Transactions on 51(6):1917 – 1937, DOI 10.1109/TIT.2005.847717

[65] Nelson S, Bakht M, Kravets R (2009) Encounter-based routing in dtns. In: INFOCOM 2009, IEEE, Rio de Janeiro, BRAZIL, pp 846–854, DOI 10.1109/INFCOM.2009.5061994

[66] Pasztor B, Musolesi M, Mascolo C (2007) Opportunistic mobile sensor data collection with scar. In: Mobile Adhoc and Sensor Systems, 2007. MASS 2007. IEEE Internatonal Conference on, Pisa, Italy, pp 1 –12, DOI 10.1109/MOBHOC.2007.4428679

[67] Pelusi L, Passarella A, Conti M (2006) Opportunistic networking: data forwarding in disconnected mobile ad hoc networks. Communications Magazine, IEEE 44(11):134–141, DOI 10.1109/MCOM.2006.248176

[68] Pentland A, Fletcher R, Hasson A (2004) Daknet: rethinking connectivity in developing nations. Computer, IEEE 37(1):78–83, DOI 10.1109/MC.2004.1260729

[69] Perkins C, Belding-Royer E, Das S (2003) Ad hoc on-demand distance vector (aodv) routing. RFC 3561, IETF Network Working Group, http://www.apps.ietf.org/rfc/rfc3561.html, accessed in December 2009

[70] Perkins CE, Bhagwat P (1994) Highly dynamic destination-sequenced distance-vector routing (dsdv) for mobile computers. In: SIGCOMM '94: Proceedings of the conference on Communications architectures, protocols and applications, ACM, New York, NY, USA, pp 234–244, DOI http://doi.acm.org/10.1145/190314.190336

[71] Piorkowski M, Sarafijanovic-Djukic N, Grossglauser M (2009) CRAWDAD data set epfl/mobility (v. 2009-02-24). Http://crawdad.cs.dartmouth.edu/epfl/mobility, accessed in December 2009

[72] Ramanathan R, Hansen R, Basu P, Rosales-Hain R, Krishnan R (2007) Prioritized epidemic routing for opportunistic networks. In: MobiOpp '07: Proceedings of the 1st international MobiSys workshop on Mobile opportunistic networking, ACM, San Juan, Puerto Rico, pp 62–66, DOI http://doi.acm.org/10.1145/1247694.1247707

[73] Rhee I, Shin M, Hong S, Lee K, Chong S (2008) On the levy-walk nature of human mobility. In: INFOCOM 2008. The 27th Conference on Computer Communications. IEEE, pp 924 –932, DOI 10.1109/INFOCOM.2008.145

[74] Scott J, Gass R, Crowcroft J, Hui P, Diot C, Chaintreau A (2009) CRAWDAD data set cambridge/haggle (v. 2009-05-29). Http://crawdad.cs.dartmouth.edu/cambridge/haggle, accessed in December 2009

[75] Sede M, Li X, Li D, Wu MY, Li M, Shu W (2008) Routing in large-scale buses ad hoc networks. In: Wireless Communications and Networking Conference, 2008. WCNC 2008. IEEE, Las Vegas, NV, USA, pp 2711–2716, DOI 10.1109/WCNC.2008.475

[76] Seth A, Kroeker D, Zaharia M, Guo S, Keshav S (2006) Low-cost communication for rural internet kiosks using mechanical backhaul. In: MobiCom '06: Proceedings of the 12th annual international conference on Mobile computing and networking, ACM, Los Angeles, CA, USA, pp 334–345, DOI http://doi.acm.org/10.1145/1161089.1161127

[77] Shah RC, Roy S, Jain S, Brunette W (2003) Data mules: modeling and analysis of a three-tier architecture for sparse sensor networks. Ad Hoc Networks 1(2-3):215 – 233, DOI DOI: 10.1016/S1570-8705(03)00003-9, sensor Network Protocols and Applications

[78] Sharma G, Mazumdar R (2004) Scaling laws for capacity and delay in wireless ad hoc networks with random mobility. In: Communications, 2004 IEEE International Conference on, Paris, France, vol 7, pp 3869 – 3873 Vol.7, DOI 10.1109/ICC.2004.1313277

[79] Sharma G, Mazumdar R, Shroff B (2007) Delay and capacity trade-offs in mobile ad hoc networks: A global perspective. Networking, IEEE/ACM Transactions on 15(5):981 –992, DOI 10.1109/TNET.2007.905154

[80] Small T, Haas ZJ (2005) Resource and performance tradeoffs in delay-tolerant wireless networks. In: WDTN '05: Proceedings of the 2005 ACM SIGCOMM workshop on Delay-tolerant networking, ACM, Philadelphia, Pennsylvania, USA, pp 260–267, DOI http://doi.acm.org/10.1145/1080139.1080144

[81] Spyropoulos T, Psounis K, Raghavendra CS (2005) Spray and wait: an efficient routing scheme for intermittently connected mobile networks. In: WDTN '05: Proceedings of the 2005 ACM SIGCOMM workshop on Delay-tolerant networking, ACM, Philadelphia, Pennsylvania, USA, pp 252–259, DOI http://doi.acm.org/10.1145/1080139.1080143

[82] Spyropoulos T, Psounis K, Raghavendra CS (2006) Performance analysis of mobility-assisted routing. In: MobiHoc '06: Proceedings of the 7th ACM international symposium on Mobile ad hoc networking and computing, ACM, Florence, Italy, pp 49–60, DOI http://doi.acm.org/10.1145/1132905.1132912

[83] Spyropoulos T, Turletti T, Obraczka K (2007) Utility-based message replication for intermittently connected heterogeneous networks. In: World of Wireless, Mobile and Multimedia Networks, 2007. WoWMoM 2007. IEEE International Symposium on a, Helsinki, Finland, pp 1–6, DOI 10.1109/WOWMOM.2007.4351693

[84] Spyropoulos T, Psounis K, Raghavendra C (2008) Efficient routing in intermittently connected mobile networks: The multiple-copy case. Networking, IEEE/ACM Transactions on 16(1):77–90, DOI 10.1109/TNET.2007.897964

[85] Spyropoulos T, Psounis K, Raghavendra C (2008) Efficient routing in intermittently connected mobile networks: The single-copy case. Networking, IEEE/ACM Transactions on 16(1):63–76, DOI 10.1109/TNET.2007.897962

[86] Tan K, Zhang Q, Zhu W (2003) Shortest path routing in partially connected ad hoc networks. In: Global Telecommunications Conference, 2003. GLOBECOM '03. IEEE, San Francisco, CA, USA, vol 2, pp 1038–1042 Vol.2, DOI 10.1109/GLOCOM.2003.1258396

[87] Vahdat A, Becker D (2000) Epidemic routing for partially-connected ad hoc networks. Duke Technical Report CS-2000-06, The Department of Computer Science, Duke University, Durham, NC

[88] Vellambi BN, Subramanian R, Fekri F, Ammar M (2007) Reliable and efficient message delivery in delay tolerant networks using rateless codes. In: MobiOpp '07: Proceedings of the 1st international MobiSys workshop on Mobile opportunistic networking, ACM, San Juan, Puerto Rico, pp 91–98, DOI http://doi.acm.org/10.1145/1247694.1247712

[89] Wang Y, Jain S, Martonosi M, Fall K (2005) Erasure-coding based routing for opportunistic networks. In: WDTN '05: Proceedings of the 2005 ACM SIGCOMM workshop on Delay-tolerant networking, ACM, Philadelphia, Pennsylvania, USA, pp 229–236, DOI http://doi.acm.org/10.1145/1080139.1080140

[90] Wang Y, Zhang P, Liu T, Sadler C, Martonosi M (2007) CRAWDAD data set princeton/zebranet (v. 2007-02-14). Http://crawdad.cs.dartmouth.edu/princeton/zebranet, accessed in December 2009

[91] Widmer J, Le Boudec JY (2005) Network coding for efficient communication in extreme networks. In: WDTN '05: Proceedings of the 2005 ACM SIGCOMM workshop on Delay-tolerant networking, ACM, Philadelphia, Pennsylvania, USA, pp 284–291, DOI http://doi.acm.org/10.1145/1080139.1080147

[92] Wu J, Yang S, Dai F (2007) Logarithmic store-carry-forward routing in mobile ad hoc networks. Parallel and Distributed Systems, IEEE Transactions on 18(6):735–748, DOI 10.1109/TPDS.2007.1061

[93] Yang Z, Wu H (2009) Featherlight information network with delay-endurable rfid support (finders). In: Sensor, Mesh and Ad Hoc Communications and Networks, 2009. SECON '09. 6th Annual IEEE Communications Society Conference on, Rome, Italy, pp 1–9, DOI 10.1109/SAHCN.2009.5168935

[94] Yoneki E, Hui P, Crowcroft J (2008) Distinct types of hubs in human dynamic networks. In: SocialNets '08: Proceedings of the 1st Workshop on Social Network Systems, ACM, Glasgow, Scotland, pp 7–12, DOI http://doi.acm.org/10.1145/1435497.1435499

[95] Zhang X, Neglia G, Kurose J, Towsley D (2006) On the benefits of random linear coding for unicast applications in disruption tolerant networks. In: Modeling and Optimization in Mobile, Ad Hoc and Wireless Networks, 2006 4th International Symposium on, Boston, Massachusetts, USA, pp 1–7

[96] Zhang X, Neglia G, Kurose J, Towsley D (2007) Performance modeling of epidemic routing. Computer Networks 51(10):2867 – 2891, DOI DOI: 10.1016/j.comnet.2006.11.028

[97] Zhang Z (2006) Routing in intermittently connected mobile ad hoc networks and delay tolerant networks: overview and challenges. Communications Surveys & Tutorials, IEEE 8(1):24–37, DOI 10.1109/COMST.2006.323440

[98] Zhao W, Ammar M, Zegura E (2004) A message ferrying approach for data delivery in sparse mobile ad hoc networks. In: MobiHoc '04: Proceedings of the 5th ACM international symposium on Mobile ad hoc networking and computing, ACM, Roppongi Hills, Tokyo, Japan, pp 187–198, DOI http://doi.acm.org/10.1145/989459.989483

[99] Zhao W, Ammar M, Zegura E (2005) Controlling the mobility of multiple data transport ferries in a delay-tolerant network. In: INFOCOM 2005. 24th Annual Joint Conference of the IEEE Computer and Communications Societies. Proceedings IEEE, Miami, FL, USA, vol 2, pp 1407–1418 vol. 2, DOI 10.1109/INFCOM.2005.1498365

Chapter 2
Opportunistic Content Distribution in Intermittently Connected Mobile Ad Hoc Networks

2.1 Overview

The last few years have witnessed an explosion of content-rich services over the Internet, and thereby content distribution itself grows into one of the most important Internet applications. Content delivery networks or content distribution networks (CDNs) first emerged to address the efficiency of content distribution over the Web for the end users, since content delivery has become important for improvement of Web performance [53]. In order to improve accessibility, decrease access delay, maximize bandwidth utilization, and maintain correctness for the users, CDN distributes content to a group of geographically dispersed cache servers. Note that these servers are located as close as possible to the users.

Recently, people are no longer pure content consumers only, but have become active content providers to share their personal contents such as digital photos, audio and videos with others over the Internet. Different technologies have been proposed to this effect. For example, people can employ the popular web-based services such as Picasa, Flickr and YouTube to publish their photos and video clips. Alternatively, a lot of peer-to-peer (P2P) based content sharing systems, e.g., Napster, Gnutella and BitTorrent are booming over the Internet to offer users a fully distributed content sharing capability without the need of a central server and its related administration. Such P2P content distribution technologies have been considered as most cost-effective solutions for these users to build their own content distribution networks [1].

The distributed nature of P2P networks even enable small computers with low bandwidth to be capable of participating in content distribution. For example , by using capabilities of users with common interest, BitTorrent distributes a content from its publisher to a number of interested users in a mesh-based manner. In BitTorrent, once the content is successfully delivered to a given peer, the peer can become another publisher to help the remaining peers to eventually receive the content. Additionally, a large content can even be divided into a number of small pieces, and once a piece of content is downloaded, it also becomes available for other peers to download. As can be seen, such a content distributing mechanism like BitTorrent leads to a flood like spreading of a content throughout the interested peers, while as more peers join the content distribution, the likelihood of a successful delivery increases. While this implies a multi-point communication mechanism, the receiver-driven strategy makes the communication different from conventional multicast approaches. Unlike the push based method of multicast where users explicitly join well known groups or topics before the intermediate routers can push contents to them, the receiver-driven P2P content delivery allows peers to solicit contents of interest from other users with the same interest. Apart from just content delivery, some more sophisticated P2P content distribution systems (e.g., Gnutella, Napster and KaZaA) also create a distributed storage medium to enable secure and efficient content publish, index, search, update, and retrieve.

Almost in parallel to content distribution, wireless technologies have also undergone a major evolution. In recent years, the wide applications of a large number of wireless handheld devices with powerful functions have been introduced to the market. Therefore, although originally developed for the Internet, such content distribution systems now transcend network boundaries (wired or wireless). Moreover, instead of the conventional cellular networks, low cost wireless connectivities such as Bluetooth and IEEE 802.11 offer the mobile devices an alternative way to communicate with each other. By utilizing such connectivities to communicate locally with their neighbors, the mobile devices can create a multi-hop network (i.e., MANET), to communicate with users and devices out of their transmission ranges. Accordingly, it will not be long before these content distribution and wireless communications will converge to enable users to create and share content on-the-fly based on spontaneously formed MANETs. Unfortunately, as mentioned earlier, end-to-end paths in MANETs may suffer intermittent connectivity due to node mobility, sporadic node density, short transmission range, and so on.

However, comparing to MANETs, ICMANs do not rely on existence of a complete end-to-end path between pair of nodes wishing to communicate with each other. By exploiting opportunistically arisen wireless links and connections instead, many delay-tolerated applications can be much more easily deployed over ICMANs without forcing all mobile nodes to form the full connected MANETs. Therefore, ICMANs offer mobile users an opportunity to reuse large amount of idle resources existing in their mobile devices to transparently and opportunistically share their personalized multimedia contents without addressing underlying connectivity. In other words, the opportunistic content sharing or content distribution mechanisms allows the contents to be effectively distributed to their potential interested mobile users without considering intermittent connectivity caused by the reasons introduced earlier. Moreover, opportunistic content distribution can also be utilized as a complementary mechanism for the Internet based content providers to publish their online contents to the corresponding mobile users by employing low-cost wireless connectivity.

However, the existing self-organized content distribution systems, which are originally developed for the Internet users, are unfeasible in the ICMANs context due to the underlying intermittent connectivity. Hence, there is growing demand for efficient architectures for deploying opportunistic content distribution systems over ICMANs, which is the primary motivation of this Chapter.

2.2 Opportunistic Content Distribution Schemes

In order to handle intermittent connectivity, mobility and the storage spaces of nodes are also utilized to facilitate opportunistic content distribution in the ICMANs context. For example, a city-wide newspaper distribution architecture is investigated in [28]. The architecture consists of several fixed Intel iMotes and some other mobile iMotes. The fixed iMotes are located at popular places as access points and content publishers generating a new content at 7 am everyday, while mobile iMotes are given to students as interested users. Based on this architecture, [28] aims to propose and evaluate schemes that distribute these contents from the access points to all mobile iMotes users. Additionally, all mobile users are assumed to belong to the same interest group, and once a content is received, it is assumed to be kept by the receiver till 7 am the next day. Different distribution schemes including selfish, collectivist, extended collaboration, top students and strangers only are evaluated in [28]. Same as previously introduced direct transmission in ICMANs, selfish content distribution only allows mobile nodes to get the content directly from the access points and downloaded content cannot be passed on to other nodes. Despite the achieved minimal delivery cost, selfish delivery involves poorest delivery performance with largest delivery delay. In contrast with selfish content distribution, collectivist based content distribution, where content is either directly downloaded from a publisher or received from other mobile nodes with the common interest, is similar to the classical content distribution mechanisms for P2P networks. By introducing extra delivery cost, considerable achievement of delivery performance is carried out by collectivist based content distribution. In addition to collectivist based distribution scheme, extended collaboration based strategy employs the capabilities of external mobile devices to relay contents to the interested users, and by exploiting the mobility and the available buffers of the nodes without the same interest, the content delivery performance of extended collaboration based strategy is further enhanced. All these content distribution schemes are illustrated in Fig. 2.1. Additionally, in top students strategy, only the N interested users that had the highest number of contacts are able to pass the contents to the other nodes, while strangers only approach only allows external devices to pass the contents.

Deploying content distribution over ICMANs is just in its early stage. Apart from the previously introduced content distribution technologies for ICMANs, the most similar approach is wireless ad hoc podcasting [29][39]. Reference [29] extensively investigates the extended collaboration based approach for wireless ad hoc podcasting by considering the restrictions of buffer spaces in mobile nodes and the limitations of wireless bandwidth. Note that podcasting has become popular for dissemination of streaming contents, especially audio streams, over the Internet, while in podcasting, once users determine which podcasts they want, they sign up for them

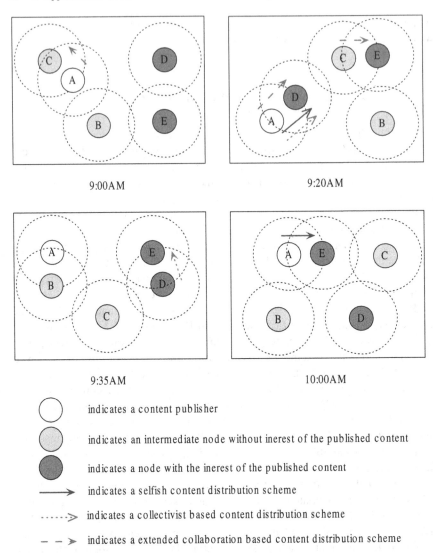

Fig. 2.1 Illustration of various content distribution schemes over ICMANs.

and have them delivered automatically whenever their devices are connected to the Internet [44]. In [29], when node *A* encounters *B*, instead of pushing its stored and cached contents to *B*, *A* allows *B* to determine which contents *B* would like to pull from *A*. In other words, *B* can make its own soliciting and caching decisions based on all received interests.

Content-based routing mechanisms using publish-subscribe [41] in the ICMANs context (e.g., [15][17][11][12]) also share some similarities with opportunistic content distributions. In a publish-subscribe system, information providers just asynchronously publish information, while information consumers place standing requests for information by issuing filters named as subscriptions. A subscription actually is a boolean function that yields true if the embedded content of information matches the consumer's interest; false otherwise. Accordingly, instead of using destination address, content-based routing allows a mobile node to route a message based on the content of this message and related subscriptions. As a result, if considering content as a general term of information[3], the major difference between content-based routing and the previous mentioned content distribution is that content-based routing for publish-subscribe enables a user to exploit subscription to request more than one contents in advance, rather than requesting a particular content. In [15][17], content is only forwarded to the nodes, which have the matching subscriptions, thereby falling into the collectivist based approaches. Moreover, [11][12] belong to extended collaboration based approaches, since it allows every node h to assign a host utility $U_{h,i}$ for each interest i it knows. $U_{h,i}$ represents the willingness of h to receive and carry messages matching the interest i. Note that if h has a subscription of h, $U_{h,i}$ is equal to 1, otherwise, it is calculated by using Kalman filters based on the context attributes such as change degree of connectivity, probability of subscriber co-location, residual energy and free buffer space. By using host utilities and forwarding utilities derived from mobility statistics, a message can eventually be pushed to its destinations.

2.3 Opportunistic Content Distribution with Extended Collaboration

As can be seen, extended collaboration based mechanisms allow each node to utilize every arising delivery opportunity for future content delivery. In other words, in order to achieve the overall content delivery, extended collaboration based opportunistic content distribution allows a node to cache and carry its uninterested contents to serve its future encounters. Therefore, it can achieve better performance over selfish based strategy and traditional collectivist based strategy. However, buffer spaces in mobile nodes and available wireless bandwidth are always limited. the main problem should be addressed is how an intermediate node can allocate its limited cache space and wireless bandwidth to the uninterested contents to achieve the overall delivery performance.

By considering the information employed by the node for making caching and forwarding decisions, once an delivery opportunity occurs, extended collaboration based mechanisms can be further classified as follows:

- *Popularity − based* approaches: In popularity-based approaches, the soliciting and caching decisions are based the popularities of all potential contents. For example, in [29], several popularity-based soliciting and caching approaches such

---

[3] In content distribution, content could be a text file, a digital photo, a MP3 file, a video-clip, or any piece of information.

as most solicited, least solicited, uniform, inverse proportional and so on are proposed. By fetching the most popular contents, most solicited strategy aims to increase the probability that the future encounters will be interested in. The goal of least solicited strategy is opposite to most solicited. It favors the contents with less popularity. This is because popular contents will be available anyway in the storage spaces of many other mobile nodes, thereby it might be more useful to cache unpopular contents in order to improve the performance of those rare contents. However, the cache space of the nodes could be filled up with the unpopular contents that never be downloaded by the future encounters. As a consequence, inverse proportional strategy is proposed to solicit and cache a content with a probability, which is inverse proportional to the popularity of the content. Moreover, with uniform soliciting and caching approach, all contents are considered equally and independent of their popularity, therefore, the uninterested contents are solicited randomly from the encounters.

- *Mobility − based* approaches: As mentioned in Chapter 1, mobility statistics are widely considered by the majority of SCF routing protocols to achieve the routing performance. Similarly, mobility statistics can also be employed for opportunistic content distribution with extended collaboration. For example, in Osmosis [20], mobility information is considered to achieve the delivery performance of opportunistic file sharing. However, in Osmosis, the response content of each content request is served by the network independently, just as a message treated by the proposed SCF routing schemes. In other words, the popularity of the content is never addressed by Osmosis.

- *Network − aware* approaches: In order to further achieve the delivery performance, mobility characteristics together with the content popularity are jointly considered by network-aware approaches. In network-aware approaches, to determine which contents can attain the caching opportunity (i.e., which contents have higher probability to be downloaded by the future encounters), the popularity of the contents, the historical mobility statistics are taken into account for content evaluation. In the following subsection, network-aware approaches will be explained in details.

2.4 Optimized Opportunistic Content Distribution Framework

This section introduces an optimized opportunistic content distribution framework for ICMANs, and accordingly proposes a Grey Relational Analysis (GRA) based decision-making mechanism for every node to solicit and cache the preferred content for the future encounters. It belongs to Extended Collaboration based mechanisms with network-aware approaches. The major material presented in this section has contributed to [35]. Note that GRA is a multi-criteria based decision-making algorithm that selects the optimal option by comparing the similarity between each option and the ideal reference vector [42]. Interested readers may refer to [42] for a detailed description of GRA algorithms, while [36] details the incorporation of GRA in formulating unicast routing approaches for ICMANs.

2.4.1 Framework Design

In the proposed content distribution framework, based on the traces related to human mobility (e.g., [28][9]), the content disseminating opportunity is assumed to only occur when any two mobile nodes encounter each other. Moreover, in order to facilitate the idea of extended collaboration [28] mentioned earlier, every node works simultaneously as a content publisher, a content consumer and a content carrier. Note that for a certain content, the content publisher is its source node and the corresponding content consumers are a group of nodes that are interested in it and have accordingly sent out their content requests. Moreover, for such a content, the content carriers are another set of nodes that themselves do not have any interest in the content, but can still cache and carry it to serve its future encounters, which may belong to the corresponding consumer group. Content carrier is a key role introduced to improve the overall delivery performance, while the storage space that is reserved for this purpose is defined as cache in this Chapter. In other words, in the proposed framework, apart from downloading its interested contents from the encounter, a node can also solicit and cache its uninterested contents from the encountering node for future dissemination based on the size of its cache. Additionally, every node is assumed to have an unlimited buffer space (i.e., private space) to store its interested contents and published contents while only a restricted cache space (public space) is reserved for the uninterested contents due to the limitation of the buffer space. This is largely because whenever a node wants to publish or download a content, it should have enough available storage space for it, while few people would like to share all buffer space with others. Note that all stored interested contents and cached uninterested contents of a node are available for all other nodes to download or solicit. Additionally, the content size itself has to be chosen so that allows at least one content can be forwarded during an encountering duration.

Since extended collaboration is utilized, performance of the proposed framework is largely determined by the efficiency of the limited cache usage, while to improve the efficiency of the corresponding soliciting and caching decision-making strategy, all content requests (i.e., the interests of content) are sprayed and stored over the network. This is because that the eventual destinations of a content are actually represented by its related content requests. Furthermore, to help other nodes update their cached contents with minimum delay, a notification of being served (i.e., anti-packet [16]) is generated and sprayed over the network whenever a mobile node receives one of its interested contents. Upon reception, an intermediate node can also remove its related content requests. In addition, since comparing to the content, the size of the content request can be ignored, every mobile node is assumed to have an unlimited space to store all received content requests. To further enhance the performance of the soliciting and caching decision-making strategy, the history of encountering information about non-random mobility in human related environments are also considered together with the content requests.

2.4.2 GRA based Soliciting and Caching Strategy (GSCS)

Due to the cache constraints as well as limited transmission capacity emerged between two encountering nodes, soliciting and caching opportunity cannot be provided to every possible content that they are not interested in. Therefore, the basic idea behind the proposed soliciting and caching strategy is to prioritize all these uninterested contents that can be cached, and accordingly solicit the more preferred contents from the encountering node. Such solicited and cached contents (i.e., more preferred contents) should have higher probability to be requested, downloaded or satisfied by the future encounters. Additionally, if the caching buffer is full, then less-preferred contents should be dropped to accommodate the preferred contents. However, the content itself cannot provide any information for prioritization since all related information comes from the interested users. For instance, the popularity of a content is determined by the number of requests it attracts (i.e., the number of users that have the common interest on the content). Conversely, many attributes affect the ability of a node to cache a content to serve a particular interested user. For example, an intermediate node may have various delivery probabilities for different users due to the non-random nature of the mobility, while a user may also have different preferences for different interested contents. Therefore, in order to rank the contents waiting for caching, the related cached requests have to be evaluated first.

There are assumed N total mobile nodes, while for a node i, its cache size is represented as B_i i.e., node i can cache up to B_i uninterested contents. Moreover, let (x,y) be any pair of two encountering nodes at a time instance t while excluding those interested contents and published contents, there are $M_{x,t}$ total distinct uninterested contents that are waiting for caching decisions from node x. To simplify the notation, M is employed in the future to represent $M_{x,t}$. Note that each such content has already been cached by node x or is available in node y for solicitation, while to simplify the formulating procedure, each of them is assigned a label m ($m \in [1,M]$) to classify itself. Furthermore, for a content m, there are L_m related cached requests, each of which is defined as $r_{m,l}, l \in L_m$. In addition, each requests has K total attributes and value of the k-th attribute of $r_{m,l}$ is represented as $r_{m,l}(k)$.

Each $r_{m,l}$ is evaluated by GRA through Grey Relational Grade (GRG), which describes the similarity between $r_{m,l}$ and a reference vector $\mathbb{R}$, which is a sequence representation of the most preferred values of all attributes for evaluation. Therefore, the less the difference is, the more preferable to be served by node x is $r_{m,l}$. The GRG of $r_{m,l}$ is carried out by utilizing

$$\Gamma_{\mathbb{R},r_{m,l}} = \sum_{k=1}^{K} \omega_k \xi_{\mathbb{R},r_{m,l}}(k) \tag{2.1}$$

where ω_k is the weight of the k-th attribute and $\xi_{\mathbb{R},r_{m,l}}(k)$ is the Grey Relational Coefficient (GRC) for $r_{m,l}(k)$, which is calculated as follows,

$$\xi_{\mathbb{R},r_{m,l}}(k) = \frac{\Delta_{min} + \zeta \Delta_{max}}{\Delta_{m,l}(k) + \zeta \Delta_{max}} \tag{2.2}$$

where

$$\Delta_{m,l}(k) = |r_{m,l}^*(k) - \mathbb{R}(k)| \tag{2.3}$$

$$\Delta_{max} = \max_{m,l} \max_{k} \Delta_{m,l}(k) \tag{2.4}$$

$$\Delta_{min} = \min_{m,l} \min_{k} \Delta_{m,l}(k) \tag{2.5}$$

and $\zeta \in [0,1]$ is employed to reduce the effect of Δ_{max}. To quantify the effects on cached request evaluation and to generate $\mathbb{R}$, $r_{m,l}(k)$ is normalized as $r_{m,l}^*(k)$ under situations: smaller-preferred and larger-preferred, where smaller-preferred is assumed for minimizing the attributes while larger-preferred governs the maximization effort. As can be seen from (2.1), the larger the GRG is, the less the difference between the corresponding request and the most preferred option is. Upon calculating the GRGs of all $r_{m,l}$, the evaluation of a content m is subsequently carried as

$$\Omega(m) = \sum_{l=1}^{L_m} \Gamma_{\mathbb{R},r_{m,l}} \tag{2.6}$$

Based on $\Omega(m)$, ranking of all possible contents is defined in an descending order i.e., the larger is the $\Omega(m)$, the more preferable is the m. Once all contents are ranked, the latest caching candidates can accordingly be determined by node x based on its cache size B_x (i.e., top B_x candidates). Comparing to the current cached contents, node x can easily find which missing contents it should solicit from its encounter y while which existing contents should be dropped to accommodate these solicited contents. However, owing to available bandwidth and content size, the encountering time between (x,y) may not allow x to solicit all its wanted contents. Therefore, based on prioritization result, the soliciting opportunities are offered to higher ranked candidates (i.e, the candidates with larger $\Omega(m)$) as long as the encountering nodes remain within the transmission range. On receiving a solicited uninterested content, if the cache is currently full, then the content with smallest $\Omega(m)$ is dropped from the cache.

Note that if the preference to each attribute (i.e., ω_k) is difficult to be given directly, the node can alteratively assign all relative preferences on a pairwise basis. For example, for a situation with three attributes, instead of directly giving the corresponding ω_k, the node can inform GRA like that: the first attribute is moderately important over the second one and very strongly important over the third one, while the second one is also moderately important over the third one. Analytical Hierarchy Process (AHP) [47], another multi-criteria based decision-making algorithm, is then applied to check the consistency and derive the exact corresponding $\{\omega_k\}$, if consistency-checking is passed. Interested readers may refer to [47] for a detailed description of AHP algorithms.

2.4.3 Considered Attributes

In order to employ GRA to evaluate every cached content request, the following attributes are considered by the proposed framework.

2.4.3.1 Expected path length (E_{path})

The first attribute considered at here is E_{path}, proposed in SEPR [52] and also utilized in [36]. As explained in the previous Chapter (e.g., Fig. 1.6), E_{path} is computed by a variation of Dijkstra algorithm and is a reasonable value to measure the distance or delivery probability between any two nodes. Based on its definition, E_{path} between any two nodes actually represents the path with highest delivery probability between them. Hence a node prefers to help serving a content request for which it has smaller E_{path} or higher delivery probability.

In fact, to calculate E_{path} for every other known node, each node in the system should maintain a table of meeting probabilities. Each node therefore has to collect the encountering information, calculate its own meeting probabilities, and generate the meeting probabilities update message with timestamp whenever it encounters any other node. The update messages are diffused over the network through flooding i.e., epidemic routing. In order to reduce the related overhead, such broadcast is limited by employing a pre-defined time-to-live (TTL) value. On receiving meeting probability update messages, each node only keeps the latest update message for every other node (ensures the latest topological view), and accordingly performs the Dijkstra algorithm to get the E_{path} of every possible content request. Moreover, by exchanging meeting probabilities, the proposed framework ensures that two encountering nodes have the same topological view before evaluating any content requests.

2.4.3.2 Differential expected path length (ΔE_{path})

E_{path} only indicates the likelihood of a certain node delivering a content to one of its interested users. Hence, even though each node can get its own optimized result based on its E_{path}, it may not be an optimized decision between the two nodes, since the decisions made by the nodes are unrelated or even conflicted. It is therefore necessary to find another attribute which can be exploited to make a collaborative decision for every content request $r_{m,l}$. Because of this, ΔE_{path} is incorporated in the proposed framework. As illustrated in Fig. 2.2, assume that at time t, pair of nodes (x,y) encounter each other, and a $r_{m,l}$ from node z is required to be served. ΔE_{path} of $r_{m,l}$ for x and y can be calculated as follows:

Fig. 2.2 E_{path} for content request $s_{m,l}$ when (x,y) encounters each other.

$$\Delta E_{path_{(x,y),z}} = E_{path_{x,z}} - E_{path_{y,z}} \qquad (2.7)$$

$$\Delta E_{path_{(y,x),z}} = E_{path_{y,z}} - E_{path_{x,z}} \qquad (2.8)$$

For every request, ΔE_{path} therefore indicates the difference between the delivery probabilities of the two encountering nodes. In fact, ΔE_{path} illustrates the relative preferences about the two encountering nodes for a $r_{m,l}$ when the system resources are limited, and helps a node to avoid making selfish decisions. From (2.7) or (2.8) and from the definition of E_{path}, it is clear that a node prefers to serve a $r_{m,l}$ for which it has a smaller ΔE_{path}. This is because smaller ΔE_{path} denotes relatively larger delivery probability to the destination.

2.4.3.3 Priority of the content request (*Prior*)

Although E_{path} and ΔE_{path} are proposed to improve the system performance from the content disseminating perspective, to address user-centric service requirements, more attributes or factors need to be utilized. To this effect, *Prior* is considered by the proposed strategy to address different service requirements from users for different contents. Here, *Prior* refers to the allocation of different priorities by the users to different $r_{m,l}$ in accordance to their own content preferences. The larger *Prior* is, more preferable the content request $r_{m,l}$ is. Hence, the proposed strategy can offer better opportunities to higher priority $r_{m,l}$ when the system resources become limited. Note that although in this Chapter only the priorities of the content requests are considered as optional attributes for application awareness, any other useful attributes from users can also easily be adopted by the proposed framework.

2.4.4 Procedure in Detail

Instead of always making decisions about which contents one node should forward to the encountering node, in the proposed framework, the node just responses the downloading or soliciting requests from its encounters. In other words, each node determines and then selects the contents it prefers to store and carry in the future by itself. This pull based mechanism helps to decrease the decision-related information exchange between the nodes. In addition, it makes it easier for mobile nodes to customize their own delivery configurations and to make reasonable delivery decisions within their abilities.

As such, the entire procedure when any two mobile nodes (x, y) encounter each other can be summarized as below:

1. Both x and y update their meeting probabilities for all past encounters and accordingly generate their latest update messages with timestamp, respectively.
2. Such latest update messages as well as all other cached update messages from other nodes are exchanged between pair (x, y) to allow them synchronize such information. On receiving such mobility update messages, each node only caches the latest update message for every other node based on the timestamp.
3. All cached content requests and anti-packets are exchanged as well. On receiving such information, the node caches all new content requests and new anti-packets, and accordingly deletes the corresponding served content requests based on the newly received anti-packets.
4. Following the exchange of these information, the summaries of all available contents are exchanged. Based on the received summaries, if there is any interested contents, they are requested from the encountering node one by one as long as pair (x, y) remain within the transmission range. Note that whenever a content request is served, the node generates an anti-packet and spreads it over the network.
5. After that, if pair (x, y) still can see each other, Dijkstra algorithm is then performed to calculate E_{path} for every possible content request, and (2.7) and (2.8) are employed by x and y to compute ΔE_{path} for each request, respectively. After that, the proposed GRA-based content soliciting and caching strategy is subsequently executed by node x and y, respectively, to carry out the soliciting sequences and the dropping sequences. Such soliciting sequences are also performed by the pair as long as they can see each other. If the buffer is currently full, the dropping sequence is performed to accommodate any solicited content.

2.4.5 Performance Evaluation

2.4.5.1 Simulation Methodology

The proposed framework with GSCS is evaluated across various cache conditions by using the OMNeT++ discrete event simulator. OMNeT++ is a C++ simulation framework, primarily applied to the network simulation [54]. It has been widely used to model different network architectures (e.g., photonic networks [10], content distribution networks [50], mobile IP networks [3], sensor networks [13][37], mobile ad hoc networks [40], vehicular networks [57], etc.), numerous routing and medium access control protocols (e.g., [55][60]), and even various web services (e.g., [6][45]).

To provide average results, the simulations were carried out independently with different seeds. In addition, the simulations are based on community based mobility models, which can reflect the real-life mobility in a campus or in a city, and has

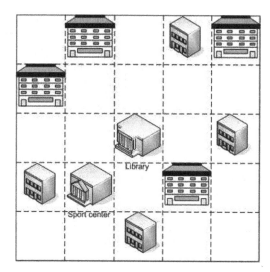

Fig. 2.3 A community based mobility model for simulation.

been extensively used in recent literature (e.g., [30]). Moreover, since the proposed framework is based on the idea of extended collaboration, to facilitate the evaluation, it is compared to another existing extended collaboration based approach [29] with most solicited and uniform strategies. As mentioned earlier, the contents with higher popularity are preferred by most solicited strategy as long as the encountering nodes can see each other, while uniform strategy just solicits the contents randomly.

Note that a total of 96 nodes are deployed in the network. Every 12 nodes form a group while the area is divided into 25 non-overlapped zones, each with an area of 200 m×200 m (shown in Fig. 2.3). There are two public-zones (*PZ1* and *PZ2*), which are randomly selected from all zones. Every 12 nodes (i.e. per group) have the same home-zone, which is also randomly picked up from all zones (excluding the public-zones). Half of these 12 nodes select *PZ1* as their preferred public-zone, while the other half select *PZ2*. Note that the public-zone is an area that has a higher probability of being visited by most of the nodes, while the home-zone has a higher probability for a particular node. This reflects a fairly populated university campus model.

The speed of the nodes is randomly chosen between [0.5 m/sec, 2 m/sec] to reflect pedestrian traffic while the pause time is uniformly distributed between [5 sec, 10 sec]. Once a node reaches its destination, the target of the next movement is determined by the related destination selection probabilities (shown in Table 2.1). Note that where the next destination should be largely depends on the present location of the node. For example, as can be seen from Table 2.1, if the node is in its home-zone, it has P_{hh} opportunity to still pick a random spot in its home-zone as the next destination, while the probability to select a spot belonging to the public-zone or elsewhere is P_{hp} and $1 - P_{hh} - P_{hp}$, respectively. It is evident from Table 2.1 that after a specific node arrival at its present location and following the pause time, it has the highest probability to pick a random spot in its home-zone.

Table 2.1 Destination selection probabilities for nodes movement

From/To	Home	Public	Elsewhere
Home	P_{hh}	P_{hp}	$1 - P_{hh} - P_{hp}$
Public	P_{ph}	P_{pp}	$1 - P_{ph} - P_{pp}$
Elsewhere	P_{eh}	P_{ep}	$1 - P_{eh} - P_{ep}$

P_{hp}	$(1 - P_{hh})P_{hp}$
P_{pp}	$(1 - P_{ph})P_{pp}$
P_{ep}	$(1 - P_{eh})P_{ep}$

P_{hh}, P_{hp}	uniform(0.7, 0.85)
P_{ph}, P_{pp}	uniform(0.7, 0.85)
P_{eh}, P_{ep}	uniform(0.8, 0.9)

2.4.5.2 Other parameters

Each node in the simulation has a transmission range of 10m to simulate the low cost wireless connectivity such as Bluetooth. As such, the available bandwidth of each node is 1Mbps. Note that in order to focus on the performance of different content distributing mechanisms, the details of the underlying wireless channels (e.g., co-channel interference and multi-path induced fading) are not simulated. This is largely because in terms of ICMAN, at most of time, communicating opportunity only occurs between any two encountering nodes, and thereby such underlying radio channel conditions should not influence the results. The simulation continues for 4 hours i.e., 14400 sec, where after the first 2000 sec, the system generates the content every 10 sec for another 10000 sec.

Pareto distribution is employed here to model the distribution of the content popularity. It is understood that highly popular contents only exhibit a small portion of the overall Internet contents [7]. In addition, similar assumption is made for the distribution of the content publishers, since most of the Internet contents are generated by a small percentage of its users, while others merely use those contents. The probability density function (pdf) of Pareto distribution is given by

$$f(x; k, x_m) = \frac{k x_m^k}{x^{k+1}} \tag{2.9}$$

For content publisher selection, $k = 1$ and $x_m = 1$ while content popularity decision follows $k = 1.5$ and $x_m = 1$. Furthermore, the consumers of a certain content are randomly selected after its popularity is determined, while to simulate the user-centric requirements, the content preference of each consumer is also randomly selected from the values of 1, 2, 3 with higher priority content request being the preferred option. Each content is assumed to have the same size (128kB is used here to simulate ringtones, wallpaper, etc. for mobile phones).

2.4.5.3 Performance Measurements

The major metrics considered in the network for evaluating performance include overall request served ratio ($Ratio_{ors}$) and the overall content caching cost ($Cost_{cnt}$)

and end-to-end request serving delay. $Ratio_{ors}$ and $Cost_{cnt}$ can be defined as follows:

$$Ratio_{ors} = \frac{\text{Total served requests}}{\text{Total requests}} \tag{2.10}$$

$$Cost_{cnt} = \frac{\text{Total content caching times}}{\text{Total served requests}} \tag{2.11}$$

In other words, $Ratio_{ors}$ defines how many content requests can be successfully served by a certain strategy, while $Cost_{cnt}$ investigates the overhead of the network on account of content caching. In addition, cumulative distribution function (CDF) of serving delay for all requests shows the distribution of the serving delay, i.e., how many requests are successfully served by considering a specific requirement of latency.

2.4.5.4 Simulation Results

Fig. 2.4 provides a comparison between GSCS and most solicited and uniform strategies in terms of $Ratio_{ors}$. Note that in GSCS-ijk, i, j and k denote the relative importance of E_{path}, ΔE_{path} and $Prior$, respectively. In addition, the caching size 0 implies the nodes only download and share the contents of their interests. In other words, the caching size 0 actually represents the collectivist based content distribution [28], since every node in the network does not have any space for the contents it is not interested in, and a content is only downloaded and stored by its interested users. Even though, "No caching" or collectivist based content distribution does not need any content soliciting and caching strategy, it is a benchmark for all extended collaboration based approaches with different soliciting and caching strategies. As expected, in terms of $Ratio_{ors}$, all extended collaboration based approaches have considerably better performance than collectivist based approach, even if a small cache size is employed by the nodes. Furthermore, such performance of all extended collaboration based approaches increases when the nodes utilize larger buffer size to cache their uninterested contents for the future encounters. Moreover, for the extended collaboration based approaches, the proposed GSCS performs better than both most solicited and uniform strategies over a range of caching sizes. The performance improvement of GSCS can be attributed to the multi-attributes based content selection by combining the content popularity with the node mobility, since GSCS allows a node to solicit and cache a content that has a higher probability to successfully satisfy and serve its future encounters.

The variation in the GSCS parametric values implies that different configurations of the relative importance of the GSCS factors can provide degree of performance differentiation. For example, as shown in Fig. 2.5, with different relative weight configurations, it is possible to achieve higher serving rates for the high priority content requests (priority 3 in this case) in resource limited environments.

In terms of $Cost_{cnt}$, the performance comparison is depicted in Fig. 2.6. While Fig. 2.4 considers successful served content requests, Fig. 2.6 investigates the con-

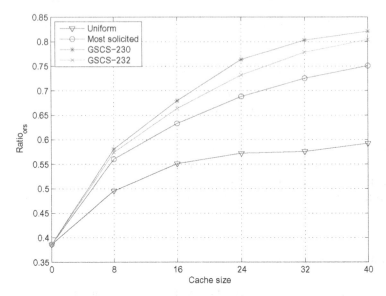

Fig. 2.4 Comparison of overall request served ratio for various strategies.

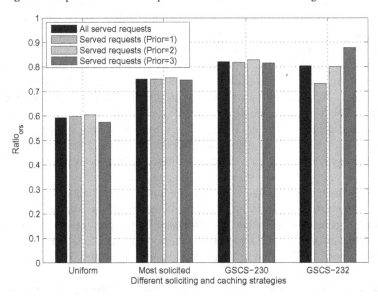

Fig. 2.5 Comparison of overall request served ratio for various strategies (cache size = 40).

tent delivery overhead of the network on account of content caching cost. As can be seen, comparing to the collectivist based approach, the delivery performance of the extended collaboration based approaches is achieved at the cost of cache utilization. Moreover, the larger is the cache space, the higher is the content delivery overhead (i.e., $Cost_{cnt}$). This is because that all these simulated extended collaboration

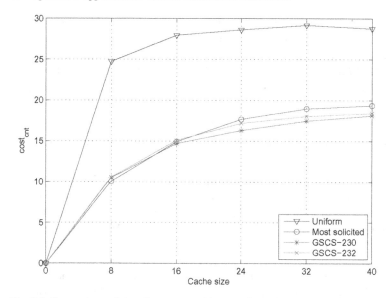

Fig. 2.6 Comparison of overall content caching cost for various caching strategies.

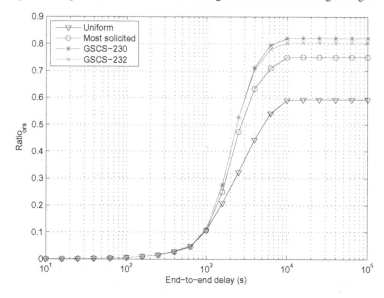

Fig. 2.7 CDFs of end-to-end request serving delay (cache size = 40).

approaches try to utilize every possible cache space for future distribution. Additionally, as shown in Fig. 2.6, despite the successful delivery ratio is achieved, in terms of $Cost_{cnt}$, GSCS just has a similar performance to the most solicited strategy, while is considerable better than the uniform strategy.

Conversely, Fig. 2.7 illustrates a CDF (cumulative distribution function) of end-to-end request serving delay for uniform, most solicited and different configurations of GSCS (cache size is 40). As is illustrated, despite GSCS offers the best performance in terms of $Ratio_{ors}$, the performance of its actual serving delay is also better than other strategies.

2.5 Cooperative Opportunistic Content Dissemination

As illustrated in previous sections, the most desirable technique to improve the efficiency of opportunistic content dissemination is to allow the intermediate nodes to solicit and cache their uninterested contents according to the available buffers so that it can better serve its potential encounters in the future [28][29]. However, in all extended collaboration based mechanisms, every intermediate node only makes its content disseminating decisions independently whenever it encounters any other node. Even though each node can get its own optimized soliciting and caching decisions for future content dissemination, it may not be an optimized decision between two encountering nodes. Hence, this section focuses on employing the idea of "cooperative" to redesign the proposed opportunistic content distribution framework presented in previous section to further enhance the delivery performance. Accordingly, a cooperative cache-based content distribution framework (CCCDF) is designed to allow the two encountering nodes to work together to carry out the cooperative soliciting and caching strategies when the resources become constrained. One significantly novel aspect of CCCDF is that two *cooperative* content disseminating strategies are introduced for the encountering nodes to make pairwise soliciting and caching decisions even under restricted resource conditions such as storage space and link bandwidth. The two strategies are presented for different motivations: CCCDF (Optimal) is an optimal strategy to maximize the overall content delivery performance while the other one, CCCDF (Max-Min), is a cooperative strategy to share the limited network resources among the contents in a Max-Min fairness manner [38] no matter how popular the content is. To carry out such cooperative strategies, a theoretical analyzing scheme is first developed to take into account all information, such as nodes mobility characteristics, content requests and buffer spaces, that are relevant for opportunistic content dissemination in ICMANs. Moreover, due to intermittent connectivity, it is difficult to get all needed information on time, thereby, a simplified local policy based on estimated information is proposed to carry out the final practical cooperative strategies. To evaluate the proposed CCCDF with different strategies, various content popularity distributions as well as two different mobility patterns are investigated in the section. Note that the major material presented in this section has contributed to [33][34].

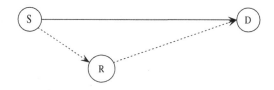

Fig. 2.8 An example of wireless cooperative communication (node R acts a relay for node S).

2.5.1 Cooperation in Wireless Networks

The idea of "cooperative" in the wireless environments is not entirely new. "Cooperative communications" on the physical layer for wireless networks are firstly gained much research interest. In cooperative communications, single antenna portable devices are allowed to share their antennas in a cooperative manner to form a virtual multiple input multiple output (MIMO) system to achieve transmission diversity [43]. As shown in Fig. 2.8, each wireless user is assumed to be not only an information sender, but also a cooperative relay to help any other user forward its information to its receiver. Accordingly, various cooperative techniques [43] such as decode and forward methods, amplify and forward methods, coded cooperation and relay selection schemes for different motivations have been introduced. For instance, in [61], energy-efficient cooperative communication is explained for wireless sensor networks based on power control and selective single-relay scheme, while a cooperative transmission scheme based on distributed space-time block coding is presented in [62] for a clustered wireless sensor network. In addition, in [23] and [21], coded cooperation is achieved through channel coding schemes. Moreover, in [32], cooperative communications in physical layer are integrated with medium access control (MAC) layer for wireless LANs to achieve throughput, delay and energy efficiency.

By taking advantage of the broadcast nature of the wireless medium and the transmission side diversity [24], in various wireless ad hoc, sensor and mesh networks with highly node density (as shown in Fig. 2.9), the cooperative multi-hop routing problem on the network layer has been recently considered in the literature as well. It has been proved that designing a cooperative routing algorithm can also lead to significant energy savings. In [22], the minimum power cooperative routing (MPCR) algorithm, a distributed cooperative routing algorithm, is introduced to choose the minimum-power route while guaranteeing the desired quality of service (QoS) parameters. By jointly investigating the problem of contention avoidance among multiple links in MAC layer and routing selection in network layer, a distributed cooperative routing scheme based on the concept of virtual node and virtual link is developed in [59] to achieve the total transmission power savings for multi-source multi-destination multi-hop wireless networks. While another energy efficient cooperative routing is considered for the distributed detection of a correlated Gaussian area in large sensor networks [51]. In addition, two different cross-layer optimization frameworks are considered in [27], one is to minimize the total power consumption for joint routing and cooperative resource allocation while the other

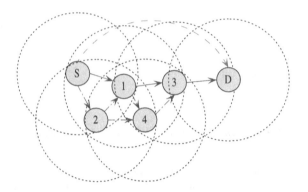

Fig. 2.9 An example of wireless cooperative routing for wireless ad hoc, sensor and mesh networks.

framework incorporates a utility-based congestion control to make a tradeoff between total power consumption minimization and the sum rate utility maximization. Apart from achieving considerable energy efficiency, [18] proposes a distributed robust cooperative routing protocol to enhance the robustness of routing against path breakage for mobile wireless sensor networks.

Cooperative diversity can also be utilized to increase transmission reliability and obtain high throughput in the face of lossy wireless links. In the literature, the cooperative routing technologies proposed for this purpose are also referred to as opportunistic protocols [5], opportunistic forwarding or opportunistic routing (OR) [31][4]. In OR, the actual node which acts as a forwarder is not known a priori by the sender, but rather is determined after the broadcast based transmission has taken place. In other words, every packet is broadcasted at first, while the exact forwarder is accordingly selected only after the set of nodes which actually received the packet is learned. By trying multiple lossy links concurrently, OR results in high expected progress per transmission. Extreme OR (ExOR) [4] is such an example that adapts cooperative diversity to standard IEEE 802.11. In ExOR, the exact forwarder is selected from the forwarder list prioritized based on expected transmission count (ETX) to the destination: the smaller the ETX, the higher the priority. Moreover, geographic random forwarding (GeRaF) [63] employs location information to choose and prioritize the forwarder set while multirate geographic opportunistic routing (MGOR) is investigated in [58] to achieves higher throughput and lower delay by incorporating the rate adaptation and candidate-selection algorithm. In addition, in MAC-Independent opportunistic routing and encoding (MORE) protocol [8], opportunistic routing is integrated with intra-flow network coding to achieve better performance than ExOR, while practical inter-flow network coding and opportunistic routing are combined together in [56] to improve the throughput of Wireless Mesh Networks (WMNs).

---

[4] The reader should distinguish the concept of opportunistic routing from the opportunistic network introduced in Chapter 1 and understand the differences between them.

2.5.2 Cooperative Cache-based Content Distribution Framework (CCCDF)

In the proposed CCCDF, the idea of "cooperative" works by allowing two encountering nodes to work jointly as a pair group to make delivery decisions for each content instead of working independently. Two decision-making strategies are investigated in this section, while the optimal cooperative disseminating strategy is studied firstly followed by the Max-Min cooperative disseminating strategy. Note that as mentioned in the previous Section, in order to adopt the idea of extended collaboration, every mobile node works simultaneously as a content publisher, a content consumer, and a content carrier, while the content dissemination only happens when any two nodes encounter each other. In addition, content disseminating decisions carried out by the strategies include interested content soliciting, uninterested content caching (i.e., uninterested content soliciting and dropping).

2.5.2.1 Optimal Cooperative Disseminating Strategy

In this subsection, the required assumptions are illustrated firstly, then the optimal cooperative caching strategy is accordingly investigated followed by the optimal cooperative interested content soliciting strategy.

N total mobile nodes are assumed in the network, while every node i has an unlimited buffer space (i.e., private space) to store its interested contents and published contents while a restricted cache space (public space) is reserved for the uninterested contents due to the limitation of the buffer space. For a node i, its cache size is represented as B_i, i.e., node i can cache up to B_i uninterested contents. Since comparing to the content, the size of content request can be ignored, every node is assumed to have an unlimited space to store all received content requests. As mentioned in the previous Section, the content size itself has to be chosen so that allows at least one content can be forwarded during an encountering duration. To eliminate regular maintaining overhead for the content requests, every content request is assumed to have an expiry time e. After this time is elapsed, if the content request still cannot be served, it is no more useful to the user and should be removed by all possible intermediate nodes from the storage. Moreover, a notification of being served (i.e., anti-packet [16]) also is generated and sprayed over the network whenever a content consumer receives its interested contents. On receiving it, an intermediate node can accordingly remove its related content request.

To formulate cooperative disseminating policies, instead of making any assumption about a particular mobility model involved, here, only the inter-encountering time between each pair of mobile nodes is assumed to follow the exponential distribution. Similar mobility assumptions are employed in [25][26][2], since most popular mobility models such as Random Waypoint, Random Direction [14] and even other community based mobility model [49] have been proved to have such a property.

Let $t_{i,j}(n)$ be the n-th encountering time between nodes i and j, then the n-th inter-encountering time $\tau_{i,j}(n)$ between nodes i and j can be defined as follows:

$$\tau_{i,j}(n) = t_{i,j}(n+1) - t_{i,j}(n)$$

By calculating the expected value of $\tau_{i,j}(n)$, $E[\tau_{i,j}]$, then the meeting rates $\lambda_{i,j}$ between nodes i and j can be carried as $1/E[\tau_{i,j}]$.

Let (x,y) be any pair of two encountering nodes at a time instance t while excluding those interested contents, there are $M_{(x,y),t}$ total distinct contents that are waiting for caching decisions from (x,y). Note that to simplify the notation, M is employed in the future to represent $M_{(x,y),t}$. In addition, each such content is uninterested by at least one of (x,y) while to simplify the formulating procedure, it is assigned a label m ($m \in [1,M]$) to classify itself. Furthermore, vectors of the potential caching content candidates of x and y at t are defined as follows:

$\mathbf{U} := (u_m)_M$ — Vector of potential caching candidates for node x, where

$$u_m = \begin{cases} 1 & \text{if content } m \text{ can be cached by node } x \\ 0 & \text{elsewhere} \end{cases}$$

$\mathbf{V} := (v_m)_M$ — Vector of potential caching candidates for node y, where

$$v_m = \begin{cases} 1 & \text{if content } m \text{ can be cached by node } y \\ 0 & \text{elsewhere} \end{cases}$$

In addition, the following information of such M contents is also assumed to be known by (x,y) at t,

$\mathbf{C} := (c_{m,n})_{M \times N}$ — Matrix of content replications across the network at t, where

$$c_{m,n} = \begin{cases} 1 & \text{if content } m \text{ is stored or cahed in node } n \\ 0 & \text{elsewhere} \end{cases}$$

$\mathbf{R} := (r_{m,n})_{M \times N}$ — Matrix of related content requests at t, where

$$r_{m,n} = \begin{cases} 1 & \text{if content } m \text{ is requested by node } n \\ 0 & \text{elsewhere} \end{cases}$$

$\mathbf{E} := (e_{m,n})_{M \times N}$ — Matrix of expiry time of related content requests

$\mathbf{G} := (g_{m,n})_{M \times N}$ — Matrix of remaining time of related content requests at t, where $\mathbf{G} = \mathbf{E} - t$

Note that all these assumptions are summarized in Table 2.2.

Given the above problem settings, a key question to answer is as follows: If (x,y) can make any cooperative content soliciting and caching decisions under the

Table 2.2 Summarized assumptions

N	Number of mobile nodes
B_i	Cache size of any node i, $i \in [1,N]$
$\Lambda := (\lambda_{i,j})_{N \times N}$	Matrix of meeting rates
t	any time instant
(x,y)	any pair of two encountering nodes x $(x,y \in [1,N])$
M	total distinct contents that can be cached by (x,y) at t
$\mathbf{U} := (u_m)_M$	Vector of potential caching candidates for node x at t
$\mathbf{V} := (v_m)_M$	Vector of potential caching candidates for node y at t
$\mathbf{C} := (c_{m,n})_{M \times N}$	Matrix of replications of M contents across the network at t
$\mathbf{R} := (r_{m,n})_{M \times N}$	Matrix of related requests of M contents at t
$\mathbf{E} := (e_{m,n})_{M \times N}$	Matrix of expiry time of $\mathbf{R}$
$\mathbf{G} := (g_{m,n})_{M \times N}$	Matrix of remaining time of $\mathbf{R}$ at t

restriction of wireless bandwidth and cache spaces, which combination of $\hat{\mathbf{U}}$ and $\hat{\mathbf{V}}$ should be considered so as to maximize overall content request served ratio. Here $\hat{\mathbf{U}}$ and $\hat{\mathbf{V}}$, the final caching decisions of nodes x and y for their possible caching candidates $\mathbf{U}$ and $\mathbf{V}$, respectively, are defined as follows:

$$\hat{\mathbf{U}} := (\hat{u}_m)_M - \text{Vector of final caching candidates for node } x, \text{ where}$$

$$\hat{u}_m = \begin{cases} 1 & \text{if content } m \text{ will be cached by node } x \\ 0 & \text{elsewhere} \end{cases}$$

$$\hat{\mathbf{V}} := (\hat{v}_m)_M - \text{Vector of final caching candidates for node } y, \text{ where}$$

$$\hat{v}_m = \begin{cases} 1 & \text{if content } m \text{ will be cached by node } y \\ 0 & \text{elsewhere} \end{cases}$$

subject to

$$\hat{u}_m \leq u_m$$
$$\hat{v}_m \leq v_m$$
$$\sum_{m=1}^{M} \hat{u}_m \equiv \min(B_x, M)$$
$$\sum_{m=1}^{M} \hat{v}_m \equiv \min(B_y, M)$$

As can be seen, the final caching contents only can be picked up by nodes x and y from the potential caching candidates $\mathbf{U}$ and $\mathbf{V}$, respectively.

Due to the exponential distribution, probability of a content request from node n for content m (i.e., $r_{m,n} \equiv 1$) that *cannot* be served by a content m holder k (i.e., $c_{m,k} \equiv 1$) in its remaining time (i.e., $g_{m,n}$) can be derived as follows:

$$p_{m,n,k} = \exp(-g_{m,n}\lambda_{k,n}) \tag{2.12}$$

Accordingly, $p_{m,n}$, the total probability of such request that *cannot* be served by the whole network in $g_{m,n}$ *without* considering pair (x,y), is represented as follows:

$$p_{m,n} = \prod_{k=1}^{N(k \neq x,y,n)} \exp(-g_{m,n}\lambda_{k,n})^{c_{m,k}}$$

$$= \exp(-g_{m,n} \sum_{k=1}^{N(k \neq x,y,n)} c_{m,k}\lambda_{k,n}) \tag{2.13}$$

Therefore, $\hat{p}_{m,n}$, the total probability of such request that *cannot* be served by the network in $g_{m,n}$ after a cooperative caching decision is made by pair (x,y), can be defined as follows,

$$\hat{p}_{m,n} = p_{m,n}p_{m,n,x}^{\hat{u}_m}p_{m,n,y}^{\hat{v}_m} \tag{2.14}$$

where

$$p_{m,n,x} = \exp(-g_{m,n}\lambda_{x,n}) \tag{2.15}$$
$$p_{m,n,y} = \exp(-g_{m,n}\lambda_{y,n}) \tag{2.16}$$

Moreover, by defining random variables $X_{m,n}$ and $\hat{X}_{m,n}$ as

$$X_{m,n} = \begin{cases} 1, & r_{m,n} \text{ can be served by the network} \\ & \text{during } g_{m,n} \text{ without considering } (x,y); \\ 0, & \text{elsewhere} \end{cases}$$

$$\hat{X}_{m,n} = \begin{cases} 1, & r_{m,n} \text{ can be served by the network} \\ & \text{during } g_{m,n} \text{ by considering } (x,y); \\ 0, & \text{elsewhere} \end{cases}$$

respectively, the expected values of $X_{m,n}$ and $\hat{X}_{m,n}$ can accordingly be represented as

$$E(X_{m,n}) = Pr(X_{m,n} = 1) = 1 - p_{m,n} \tag{2.17}$$

$$E(\hat{X}_{m,n}) = Pr(\hat{X}_{m,n} = 1) = 1 - \hat{p}_{m,n} \tag{2.18}$$

since $X_{m,n}$ and $\hat{X}_{m,n}$ follow Bernoulli distribution. Accordingly, for all independent requests related with m, by defining random variable $\hat{X}_m$ as total number of requests that can be served by the system after the cooperative caching decision is made by pair (x,y), its expected value can be represented as follows:

$$E[\hat{X}_m] = \sum_{n=1}^{N} r_{m,n}(1 - \hat{p}_{m,n})$$

$$= \sum_{n=1}^{N} r_{m,n}(1 - p_{m,n}P_{m,n,x}^{\hat{u}_m}P_{m,n,y}^{\hat{v}_m}) \qquad (2.19)$$

Moreover, if a random variable $\hat{X}$ is defined as total number of successfully served content requests for all known content candidates after the cooperative caching decision is made by pair (x,y), the expected value of $\hat{X}$ can be carried out as below:

$$E[\hat{X}] = \sum_{m=1}^{M} E[\hat{X}_m]$$

$$= \sum_{m=1}^{M} \sum_{n=1}^{N} r_{m,n}(1 - p_{m,n}P_{m,n,x}^{\hat{u}_m}P_{m,n,y}^{\hat{v}_m}) \qquad (2.20)$$

Similarly, for all independent requests related with m, if random variable X_m is defined as total number of requests that can be served by the system without help from pair (x,y), its expected value can be represented as follows:

$$E[X_m] = \sum_{n=1}^{N} r_{m,n}(1 - p_{m,n}) \qquad (2.21)$$

while the expected value of X, total number of successfully served content requests for all known content candidates without help from pair (x,y), can subsequently be represented as below:

$$E[X] = \sum_{m=1}^{M} \sum_{n=1}^{N} r_{m,n}(1 - p_{m,n}) \qquad (2.22)$$

Note that in order to maximize overall content request served ratio, $E[\hat{X}]$ should be maximized by pair (x,y), hence the optimal content caching strategy is to find $\hat{U}$ and $\hat{V}$ satisfying the following criteria:

$$(\hat{U}, \hat{V}) = \arg\max[E[\hat{X}]] \qquad (2.23)$$

Moreover, as can be seen from (2.22), $E[X]$ is a constant value. Therefore, (2.23) can alternatively be expressed as follows:

$$(\hat{U}, \hat{V}) = \arg\max[E[\hat{X}] - E[X]] \qquad (2.24)$$

Let $\Delta = E[\hat{X}] - E[X]$, therefore,

$$\Delta = \sum_{m=1}^{M} \Delta_m$$

$$= \sum_{m=1}^{M} \sum_{n=1}^{N} r_{m,n} P_{m,n} (1 - p_{m,n,x}^{\hat{u}_m} p_{m,n,y}^{\hat{v}_m}) \tag{2.25}$$

As can be seen from (2.25), $\Delta_m = \sum_{n=1}^{N} r_{m,n} P_{m,n}(1 - p_{m,n,x}^{\hat{u}_m} p_{m,n,y}^{\hat{v}_m})$ is a non-negative value representing a margin expected value of total served requests that pair (x,y) can offer content m. Therefore, to achieve (2.23) or (2.24), $\hat{U}$ and $\hat{V}$ can be obtained by always offering the cache space to the content with the maximized Δ_m until the cache space in pair (x,y) is full.

As a consequence, the system calculates the margin expected value of total served requests for every content m by assuming only one buffer space in nodes x or y can be offered firstly, i.e.,

$$\Delta_m(x) = \sum_{n=1}^{N} r_{m,n} P_{m,n} (1 - p_{m,n,x}) \tag{2.26}$$

$$\Delta_m(y) = \sum_{n=1}^{N} r_{m,n} P_{m,n} (1 - p_{m,n,y}) \tag{2.27}$$

Then, if $\Delta_m(x)$ is greater than $\Delta_m(y)$, $\Delta_m(y|x)$ is computed instead of $\Delta_m(y)$, i.e.,

$$\Delta_m(y) = \Delta_m(y|x) = \sum_{n=1}^{N} r_{m,n} P_{m,n} P_{m,n,x} (1 - p_{m,n,y}) \tag{2.28}$$

otherwise,

$$\Delta_m(x) = \Delta_m(x|y) = \sum_{n=1}^{N} r_{m,n} P_{m,n} P_{m,n,y} (1 - p_{m,n,x}) \tag{2.29}$$

Note that, if a content m can be only cached by node x (which means that node y is interested in m and has downloaded it), we only calculate $\Delta_m(x)$ from (2.29), and vice versa. After that, by prioritizing all $\Delta_m(x)$ and $\Delta_m(y)$ in a descending order, we can finally find the optimal decision for $\hat{U}$ and $\hat{V}$. By comparing $\hat{U}$ and $\hat{V}$ to currently cached content candidates, node x and y can learn which contents to solicit from each other while which contents to drop from the cache. Obviously, if $M \leq \min(B_x, B_y)$, the optimal policy is to cache all uninterested content candidates in both nodes x and y.

As can be seen from (2.26) - (2.29), in order to carry out the optimal cooperative caching strategy, the latest information such as Λ and C are required. However, because of intermittent connectivity, it is difficult for the intermediate nodes to collect all these latest information on time in the ICMANs context. Moreover, to maintain Λ and C, each node also has to collect the latest information about its meeting rates with all other nodes as well as all its stored contents and accordingly spreads them over the network regularly. Such maintaining overhead dominates control overhead as the network size and the buffer sizes increase. Therefore, the equations expressed earlier have to be simplified.

Note that if the average meeting rate $\tilde{\lambda}_i$ of each node i to all other nodes, i.e.,

$$\tilde{\lambda}_i = \frac{1}{N-1} \sum_{j=1}^{N(j\neq i)} \lambda_{i,j} \tag{2.30}$$

is employed instead, $p_{m,n}$ can be approximated as follows,

$$p_{m,n} \approx \tilde{p}_{m,n} = \exp(-g_{m,n}\tilde{\lambda}_n l_m) \tag{2.31}$$

where l_m is the existing number of stored content m over the network excluding pair (x,y). Accordingly, (2.26) - (2.29) can be represented as follows:

$$\Delta_m(x) \approx \sum_{n=1}^{N} r_{m,n}\tilde{p}_{m,n}(1 - p_{m,n,x}) \tag{2.32}$$

$$\Delta_m(y) \approx \sum_{n=1}^{N} r_{m,n}\tilde{p}_{m,n}(1 - p_{m,n,y}) \tag{2.33}$$

$$\Delta_m(y) = \Delta_m(y|x) \approx \sum_{n=1}^{N} r_{m,n}\tilde{p}_{m,n}p_{m,n,x}(1 - p_{m,n,y}) \tag{2.34}$$

$$\Delta_m(x) = \Delta_m(x|y) \approx \sum_{n=1}^{N} r_{m,n}\tilde{p}_{m,n}p_{m,n,y}(1 - p_{m,n,x}) \tag{2.35}$$

Now every node i just needs to flood its $\tilde{\lambda}_i$ across the network regularly to enable all other nodes maintain $\tilde{\Lambda}$ while instead of collecting $\mathbf{C}$, every node only has to estimate l_m for each possible content m. Note that for every cached content, an intermediate node knows that there is always a publisher. Besides, an intermediate node also can learn from the cached anti-packets that for every cached content, up to now at least how many corresponding consumers in the system have permanently stored it. Combining these two together actually is the lower bound of l_m, therefore to simplify our process, we just use this value to approximate the actual l_m.

The former optimal cooperative content caching and dropping policy is achieved without addressing the limitation of link bandwidth or the encountering duration. However, owing to available bandwidth and the content size, the encountering time between pair (x,y) may not allow transfer of all solicited contents. Therefore, based on prioritization result, the soliciting opportunities is offered to higher ranked candidates (i.e, the candidates with larger $\Delta_m(x)$ or $\Delta_m(y)$) as long as the encountering nodes remain within the transmission range. On receiving a solicited uninterested content, if the cache is currently full, then the content with smallest $\Delta_m(x)$ or $\Delta_m(y)$ is dropped from the cache.

Actually, the encountering time between pair (x,y) may be also not long enough to finish soliciting their interested contents. Therefore, similar ranking procedure can be utilized for such interested contents as well. The margin expected value of total served requests of all interested contents for node x, which can be downloaded from node y, are computed based on (2.32) while such values of interested contents for node y are carried out according to (2.33). After that, the soliciting opportunity is assigned to the interested contents with higher ranked results.

2.5.2.2 Max-Min Fairness Cooperative Disseminating Strategy

In the optimal cooperative disseminating strategy, resources (wireless link, private and public spaces) are allocated to maximize overall content request served probability. However, such strategy cannot ensure that the available network resources are shared among the contents in a fair manner. For example, the requests for a popular content may easily get more downloading opportunities since the content are available anyway in the private spaces of more users. As a result, a Max-Min fairness based cooperative disseminating strategy is proposed here to improve the delivery opportunity for the unpopular contents. The Max-Min fairness [38][46][19] is a well-known fairness criterion that has been applied to resource managements in different areas such as rate control, flow control, wireless scheduling and multicast protocols. It gives the most poorly treated user the largest possible resources, while not unnecessarily wasting any resources. Here a cache space and wireless link allocation is said to be Max-Min fairness when it is impossible to increase the expected serving probability of a content without reducing the probability of another content with a worse served performance. In other words, it is a Max-Min fairness allocation when: Firstly, the expected serving probability of the worst served content is maximized; secondly, the expected serving probability of the second worst served content is maximized, etc.

Based on previously defined $E[X_m]$, the expected served ratio of a content m ($R_{srvd}(m)$) without considering (x,y) can be defined as follows:

$$R_{srvd}(m) = \frac{q(m)+E[X_m]}{q(m)+\sum_{n=1}^{N} r_{m,n}}$$
$$\approx \frac{q(m)+\sum_{n=1}^{N} r_{m,n}(1-\tilde{p}_{m,n})}{q(m)+\sum_{n=1}^{N} r_{m,n}} \qquad (2.36)$$

where $q(m)$ indicates the number of already served requests of content m while $\tilde{p}_{m,n}$ is utilized to approximate $p_{m,n}$. Accordingly, the expected served ratio of a content m by only considering node x ($R_{srvd-x}(m)$) or y ($R_{srvd-y}(m)$) can be defined as follows:

$$R_{srvd-x}(m) \approx \frac{q(m)+\sum_{n=1}^{N} r_{m,n}(1-\tilde{p}_{m,n}p_{m,n,x})}{q(m)+\sum_{n=1}^{N} r_{m,n}} \qquad (2.37)$$

$$R_{srvd-y}(m) \approx \frac{q(m) + \sum_{n=1}^{N} r_{m,n}(1 - \tilde{p}_{m,n} p_{m,n,y})}{q(m) + \sum_{n=1}^{N} r_{m,n}} \qquad (2.38)$$

Based on (2.36) - (2.38), the Max-Min fairness based cooperative caching policy can eventually be carried out as follows: All caching candidates are ranked based on their $R_{srvd}(m)$ in an ascending order firstly, while the candidate with smallest value will be offered a caching opportunity by (x,y). If the related $R_{srvd-x}(m)$ is greater than $R_{srvd-y}(m)$, such an opportunity comes from x; otherwise, it is allocated from y. After that, this candidate is updated with a new expected value and all contents are ranked again. Such process is continued until all caching spaces are assigned out or all content candidates get the cache spaces that they want. Note that if the latest and the lowest ranked candidate only needs a caching space from one node, it is offered by the corresponding node directly and is removed from the ranking procedure. Moreover, if necessary, such cache space allocation sequence also is the wireless access sequence while its reverse sequence represents the dropping sequence when the cache space is full. A similar procedure can be utilized for the interested contents as well.

2.5.3 Procedure in Detail

Comparing to the framework proposed in the previous Section, where each node just makes its independent soliciting and dropping decision, even though all nodes work together to collect all required information, CCCDF employs the idea of "cooperative". Therefore, the previous proposed procedure cannot be used directly here. Hence, in this subsection, the procedure of CCCDF is explained in detail.

Note that in the proposed CCCDF, only one of CCCDF (Optimal) and CCCDF (Max-Min) is needed. Moreover, once it is determined, only one node has to make the cooperative decisions when (x,y) encounter each other. If we assume $x < y$, then we let node x make such cooperative decisions for both of them, therefore the procedure can be summarized as below:

1. Both nodes update their λ for all past encounters and accordingly calculate and generate the latest update message of $\tilde{\lambda}_x$ and $\tilde{\lambda}_y$ with timestamp, respectively.
2. To assist content soliciting procedure, such latest average meeting rate update message as well as all other cached update messages for $\tilde{\lambda}_k, k \in [1, N]$ are exchanged between (x,y) to allow them synchronize such information. On receiving such mobility update messages, each node only caches the latest $\tilde{\lambda}_k$ for every other node based on the timestamp. Moreover, all cached content requests and anti-packets are exchanged as well. On receiving such information, the node only caches all new requests and new anti-packets, and deletes the cached requests, which have already been served, based on newly received anti-packets.
3. After that, node y forwards the summaries of its available contents and its latest meeting rates $\lambda_{y,k}, k \in [1, N]$ to node x. Upon reception, node x accordingly employs the soliciting strategy for the interested contents to find the correspond-

ing soliciting sequence for both nodes. Such soliciting sequence is performed as long as (x,y) remain within the transmission range. Note that whenever a content request is served, the node generates an anti-packet and spreads it over the network.

4. If (x,y) still can see each other, the caching strategy is then executed to get the uninterested contents soliciting sequence as well as dropping sequence for nodes x and y, respectively. Such soliciting sequence also is performed by (x,y) as long as they can see each other, while if the buffer is currently full, the dropping sequence is performed to accommodate any solicited content.

2.5.4 Performance Evaluation

2.5.4.1 Simulation Methodology

The goal of the evaluation in this Section is to show that the proposed CCCDF can get a better performance compared to the existing work when the cache size is restricted. As mentioned earlier, cache is a portion of the storage space reserved only for the uninterested contents. Note that the same simulator (i.e., the OMNeT++ discrete event simulator) is also employed in this Section to evaluate the performance of CCCDF across various cache conditions. Furthermore, instead of using a specific mobility model such as community-based mobility model used earlier, the simulation at here is based on two generalized mobility patterns, a synthetic one based on the exponential distribution and one real-world mobility trace collected in Cambridge, UK [28]. Note that the encountering times between mobile nodes in the simulation are not known *a priori*. All information is learned during the simulation. In addition, Bluetooth is also employed at here, therefore, the available bandwidth of each node is set as 1Mbps. Moreover, a smaller size (i.e., the size of 64 kB) is assumed for each content here.

Since CCCDF is also based on the idea of external collaboration, to facilitate such evaluation, two proposed soliciting and caching strategies are compared to other strategies such as most solicited and uniform strategies in [29] and GSCS. As mentioned previously, the contents with higher popularity are preferred by most solicited strategy as long as the encountering nodes can see each other, while uniform strategy just solicits the contents randomly. Moreover, GSCS ranks all possible content candidates by evaluating all related content requests based on history encountering information and the content preferences of the consumers, and accordingly solicits and caches higher ranked contents. Note that such content request evaluations are carried out by employing GRA, while the priorities assigned to both Expected path length, Differential expected path length and Priority of the content request at here are 2, 3, and 0, respectively.

Table 2.3 Summarized parameters

Number of mobile nodes	64
Pair of nodes encountering probability	0.4
$E[\tau_{i,j}]$	$N(1500sec, 300sec)$
Encountering duration	$N(12sec, 4sec)$
Simulation duration	$4hours$
Warming up duration	first hour
Content generating frequency	every 6 sec
Content generating duration	9600 sec
Request expiry time	1200 sec

2.5.4.2 Performance Measurements

Apart from the previous introduced overall request served ratio ($Ratio_{ors}$ in (2.10)), overall content caching cost ($Cost_{cnt}$ in (2.11)) and end-to-end request serving delay, two other major metrics including overall content served ratio ($Ratio_{ocs}$) and average requests per content served ratio ($Ratio_{arcs}$) for evaluating performance are considered at here. $Ratio_{ocs}$ and $Ratio_{arcs}$ are defined as follows:

$$Ratio_{ocs} = \frac{\text{Total contents with served requests}}{\text{Total contents}} \qquad (2.39)$$

$$Ratio_{arcs} = \frac{1}{\text{Total contents}} \sum \frac{\text{Total srvd requests per content}}{\text{Total requests per content}} \qquad (2.40)$$

In other words, $Rate_{ocs}$ and $Rate_{arcs}$ investigate how fairness a certain strategy can achieve.

2.5.4.3 Results from synthetic mobility model

In this subsection, a synthetic mobility model is employed to compare the performance of CCCDF (Optimal) and CCCDF (Max-Min) to GSCS, most solicited and uniform strategies. In such mobility model, the inter-meeting time between any pair of encountering nodes is represented as an exponential distribution as explained earlier. We randomly determine which pair of nodes has an opportunity to encounter each other during the whole simulation and such probability is set as 0.4. Once two nodes can encounter each other, then the $E[\tau_{i,j}]$ (i.e., $1/\lambda_{i,j}$) of its inter-encountering time follows a Gaussian distribution with $N(1500 \text{ sec}, 300 \text{ sec})$. Moreover, every encountering duration follows another Gaussian distribution with $N(12 \text{ sec}, 4 \text{ sec})$. Note that here the simulation considers 64 nodes. The simulation continues for 4 hours, i.e., 14400 sec where after the first 1 hour, the system generates the content and its related content requests every 6 sec for another 9600 sec. Furthermore, the expiry time of every content request is set as 1200 sec.

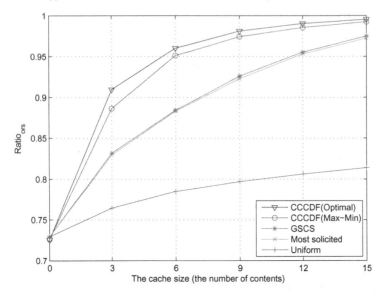

Fig. 2.10 Comparison of overall request served ratio for various strategies (uniform distribution).

Firstly, the uniform distribution is utilized to simulate the distribution of the content publishers and the content popularity. In other words, for each generated content, the publisher ID is randomly selected from all 64 available nodes while its popularity (i.e., the number of corresponding consumers) are randomly picked up from [1,24]. Note that the consumer ID of a certain content are randomly selected after its publisher ID and popularity is determined.

Fig. 2.10 provides the performance of $Ratio_{ors}$ when the cache size is varied. Cache size 0 implies the nodes only download and share the contents of their interests (i.e., collectivist). Even though, "No caching" is not a real content caching strategy, it can be a benchmark for all real caching strategies. As expected, all caching strategies have considerably better performance than "No caching", even if a small cache size is employed. Furthermore, such performance of the caching strategies increases when the nodes utilize larger buffer size to cache the uninterested contents for the future encounters. From these plots, it can be seen that CCCDF (Optimal) gives the best performance across all cache size conditions. In other words, comparing to all other caching strategies, CCCDF (Optimal) always can use less cache size to offer the same request serving ratio. Note that the performance improvement of CCCDF (Optimal) could be deducted from its optimal cooperative disseminating strategy based on the characteristics of the mobility as well as content requests. In addition, as can be seen from Fig. 2.10, performance of CCCDF (Max-Min) is close to CCCDF (Optimal) while also is much better than other caching strategies when the content popularity is uniform distributed.

While Fig. 2.10 considers overall request served ratio, Fig. 2.11 and Fig. 2.12 investigate the overall content served ratio and the average requests served ratio per

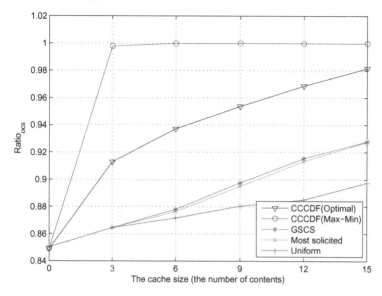

Fig. 2.11 Comparison of overall content served ratio for various strategies (uniform distribution).

content over different cache sizes, respectively. As can be seen, by employing Max-Min fairness policy, CCCDF (Max-Min) can give more opportunities to the contents with less interests and accordingly serve more requests for such unpopular contents. For example, when cache size is 3, CCCDF (Max-Min) strategy can serve at least a request for most of contents no matter how popular it is while the average requests served ratio per content is 85%. And comparing to CCCDF (Optimal), both of them are around 8% higher. Moreover, as can be seen from Fig. 2.11 and Fig. 2.12, both CCCDF strategies offers much better fairness performance than other strategies in such a scenario with uniform distributed content popularity.

Fig. 2.13 investigates the content caching overhead of the network on account of uninterested content caching i.e., $Cost_{cnt}$ over different cache sizes. As can be seen, CCCDF strategies just use similar caching overhead compared to GSCS and most solicited to achieve considerably better content delivery performance. Conversely, Fig. 2.14 illustrates CDFs of request serving delay for all caching strategies when cache size is 6. Fig. 2.14 shows that even though CCCDF (Optimal) can serve more content requests before their expected deadlines, the performance of its actual end-to-end serving delay is also better than other strategies.

The second distribution considered in the simulation for the content popularity and the content publishers is Pareto distribution, which has been explained previously. Note that for content publisher selection, $k = 1$ and $x_m = 1$ is considered while content popularity decision follows $k = 2$ and $x_m = 1$.

The performance comparisons between CCCDF and other mechanisms are provided in Fig. 2.15 - 2.19. As can be seen, in terms of R_{ors}, CCCDF (Optimal) offers the best performance among all caching strategies when the cache space is restricted.

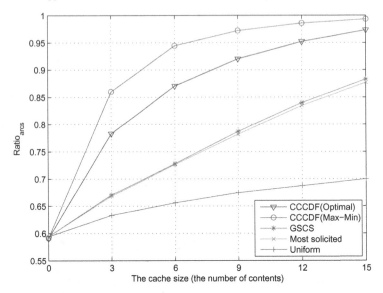

Fig. 2.12 Comparison of average requests per content served ratio for various strategies (uniform distribution).

Moreover, comparing with the previous scenario, performance of CCCDF (Optimal) is over other strategies more considerably. For instance, when cache size is 3, the performance of CCCDF (Optimal) is more than 15% better than all other strategies. The reason of such considerable achievement is behind (2.32) - (2.35).

As can be seen, when the majority of the contents are not popular (i.e., only have relatively few consumers), Δ_m of different contents are more determined by meeting statistics rather than content popularity. This can also be proved by the performance of GSCS, since comparing to the previous scenario, GSCS can also achieve substantial better performance over most solicited strategy when content popularity follows a Pareto distribution.

However, performance of CCCDF (Max-Min) is degraded when the content popularity follows Pareto distribution. This is because in such a scenario, the majority of the published contents only have few interested users while for an unpopular content, its consumers has less opportunity to download from the users having the same interest, therefore, the cache space becomes more essential to serve such content request. Accordingly, sharing the limited caching opportunities in a fair manner based on the Max-Min fairness policy causes that few of such content can get enough caching resources to successfully serve their interested consumers, thereby degrades its delivery performance. It also is proved by Fig. 2.16, Fig 2.17 and Fig 2.19 where the performance of CCCDF (Max-Min) is degraded as well. In general, for such Pareto distributed case, the overall performance of the CCCDF (Optimal) is better than the CCCDF (Max-Min), while for the previous uniform distributed scenario, the CCCDF (Max-Min) performs better than the CCCDF (Optimal).

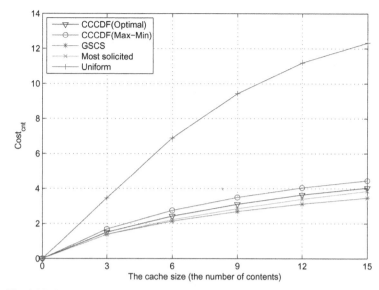

Fig. 2.13 Comparison of overall content caching cost for various caching strategies (uniform distribution).

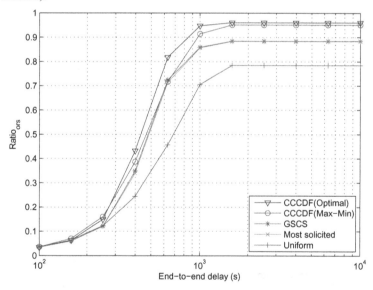

Fig. 2.14 CDFs of end-to-end request serving delay (cache size = 6) (uniform distribution).

2.5.4.4 Results based on real-world mobility trace

As mentioned earlier, the real-world mobility trace is collected from an experiment conducted in the city of Cambridge. The whole experiment focuses on tracking con-

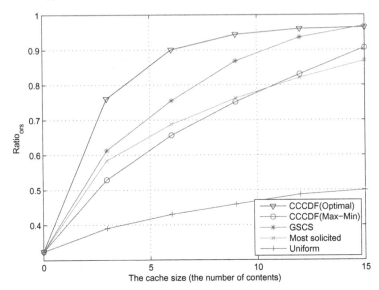

Fig. 2.15 Comparison of overall request served ratio for various strategies (Pareto($k = 2, x_m = 1$)).

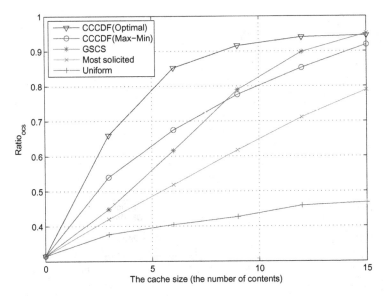

Fig. 2.16 Comparison of overall content served ratio for various strategies (Pareto($k = 2, x_m = 1$)).

tracts between various mobile nodes and between mobile nodes and different fixed nodes as well. Mobile users in the experiment are mainly from University of Cambridge who were requested to carry Intel iMote devices with them at all times for the duration of the experiment. In addition to this, a number of stationary nodes are deployed in various popular locations such as market places, shopping centers and

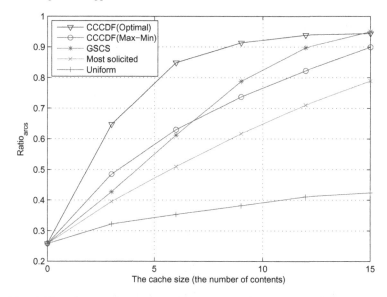

Fig. 2.17 Comparison of average requests per content served ratio for various strategies (Pareto($k = 2, x_m = 1$)).

pubs. Note that such Intel iMotes can log contacts with other Bluetooth integrated devices. The experiment was conducted from October 28th, 2005, 9:55:32 (GMT) to December 21st, 2005, 13:00 (GMT). While due to various hardware problems and the loss of some of the deployed iMotes, measurement data were gathered from 36 mobile participants and 18 fixed locations. For a detailed description of the experiment, interested readers may refer to [28] and note that the mobility trace from the experiment can be downloaded from [48].

Note that only content distribution among mobile users is considered, therefore only traces of 36 mobile participants are employed. Furthermore, by considering the cooperative case, " node A sees node B" is assumed to implies "node B sees node A" as well, vice versa. In addition, because of the average lifetime of mobile nodes is around 10 days, the simulation only continues for first 12 days, where after the first 24 hours, the system generates the content with its related requests every half an hour for another 8 days. For each content request, the expiry time is set as 3 days (i.e., each request has to be served by the network in 72 hours after it is generated). Moreover, Pareto distribution also is employed here while ($k = 1, x_m = 1$) and ($k = 1.5, x_m = 1$) are considered for for content publisher selection and popularity decision, respectively.

Fig. 2.20 provides a comparison between them in terms of R_{ors}. From these plots, it can be seen that CCCDF (Optimal) gives the best performance across all cache size conditions. For example, for cache size 4, CCCDF (Optimal)'s overall request served ratio is around 5% higher than CCCDF (Max-Min), GSCS and most solicited while 12% higher than uniform strategy. Moreover, as explained earlier, when the content popularity follows Pareto($k = 1.5, x_m = 1$), the fairness perfor-

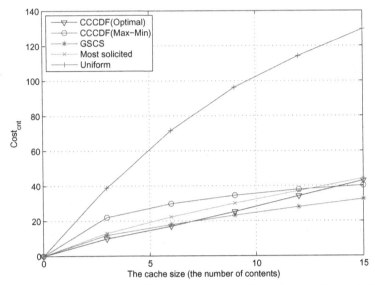

Fig. 2.18 Comparison of overall content caching cost for various caching strategies (Pareto($k = 2, x_m = 1$)).

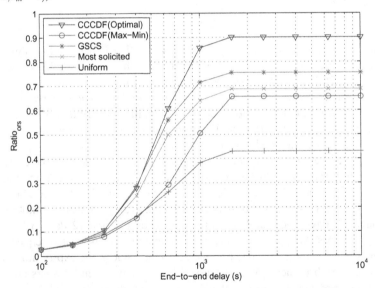

Fig. 2.19 CDFs of end-to-end request serving delay (cache size = 9) (Pareto($k = 2, x_m = 1$)).

mance (i.e., $Ratio_{ocs}$ and $Ratio_{arcs}$) of CCCDF (Optimal) strategy also is close to or even better than the CCCDF (Max-Min) strategy for the real-world mobility (shown in Fig. 2.21 and Fig. 2.22). In addition, the content caching overhead of the network (i.e., $Cost_{cnt}$) over different cache sizes is investigated in Fig. 2.23. As can

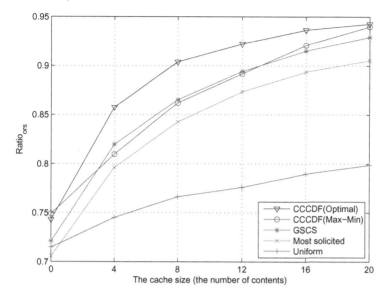

Fig. 2.20 Comparison of overall request served ratio for various strategies (Pareto($k = 1.5, x_m = 1$)).

be seen, CCCDF strategies offer similar caching overhead compared to GSCS and most Solicited. Fig. 2.24 also shows that even though CCCDF (Optimal) can serve more content requests before their expected deadlines, the performance of its actual end-to-end serving delay is also better than other strategies.

2.6 Summary and Conclusions

This Chapter first briefly introduced the current efforts of opportunistic content distribution mechanisms for ICMANs. After that, an extended collaboration based content delivery framework with a novel content soliciting and caching strategy (GSCS) by jointly considering multi-attributes (e.g., content popularity, mobility statistics of nodes and preferences of users) was described in this Chapter. Simulation results indicated that the proposed framework can work well in realistic scenarios. In this Chapter, another new framework based on the idea of "cooperative" was also proposed for content dissemination over ICMANs. Unlike other protocols where every intermediate node made its content disseminating decisions without considering its encounters, CCCDF considered two encountering nodes as a pair and accordingly proposed an analytical model based on the global information to carry out the content disseminating decisions even under restricted limited resource scenarios. Moreover, to handle intermittent connectivity, a simplified local policy was illustrated to approximate the proposed cooperative strategies. Experimental results

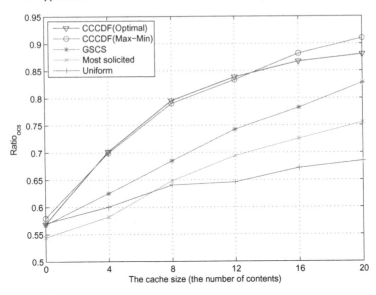

Fig. 2.21 Comparison of overall content served ratio for various strategies (Pareto($k = 1.5, x_m = 1$)).

indicated an enhanced performance of the proposed framework compared with the existing mechanisms.

References

[1] Androutsellis-Theotokis S, Spinellis D (2004) A survey of peer-to-peer content distribution technologies. ACM Comput Surv 36(4):335–371, DOI http://doi.acm.org/10.1145/1041680.1041681

[2] Balasubramanian A, Levine B, Venkataramani A (2007) Dtn routing as a resource allocation problem. In: SIGCOMM '07: Proceedings of the 2007 conference on Applications, technologies, architectures, and protocols for computer communications, ACM, Kyoto, Japan, pp 373–384, DOI http://doi.acm.org/10.1145/1282380.1282422

[3] Benmammar B, Jrad Z, Krief F (2009) Qos management in mobile ip networks using a terminal assistant. Int J Netw Manag 19(1):1–24, DOI http://dx.doi.org/10.1002/nem.684

[4] Biswas S, Morris R (2005) Exor: opportunistic multi-hop routing for wireless networks. In: SIGCOMM '05: Proceedings of the 2005 conference on Applications, technologies, architectures, and protocols for computer communications, ACM, Philadelphia, Pennsylvania, USA, pp 133–144, DOI http://doi.acm.org/10.1145/1080091.1080108

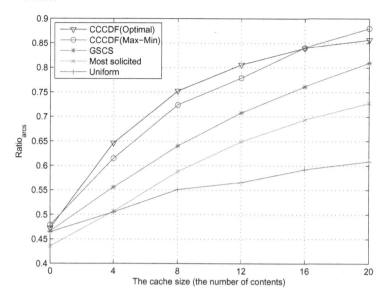

Fig. 2.22 Comparison of average requests per content served ratio for various strategies (Pareto($k = 1.5, x_m = 1$)).

[5] Campista M, Esposito P, Moraes I, Costa L, Duarte O, Passos D, de Albuquerque C, Saade D, Rubinstein M (2008) Routing metrics and protocols for wireless mesh networks. Network, IEEE 22(1):6–12, DOI 10.1109/MNET.2008.4435897

[6] Canali C, Colajanni M, Lancellotti R (2010) Resource management strategies for the mobile web. Mob Netw Appl 15(2):237–252, DOI http://dx.doi.org/10.1007/s11036-009-0186-1

[7] Cha M, Kwak H, Rodriguez P, Ahn YY, Moon S (2009) Analyzing the video popularity characteristics of large-scale user generated content systems. Networking, IEEE/ACM Transactions on 17(5):1357–1370, DOI 10.1109/TNET.2008.2011358

[8] Chachulski S, Jennings M, Katti S, Katabi D (2007) Trading structure for randomness in wireless opportunistic routing. In: SIGCOMM '07: Proceedings of the 2007 conference on Applications, technologies, architectures, and protocols for computer communications, ACM, Kyoto, Japan, pp 169–180, DOI http://doi.acm.org/10.1145/1282380.1282400

[9] Chaintreau A, Hui P, Crowcroft J, Diot C, Gass R, Scott J (2007) Impact of human mobility on opportunistic forwarding algorithms. Mobile Computing, IEEE Transactions on 6(6):606–620, DOI 10.1109/TMC.2007.1060

[10] Chan J, Hendry G, Biberman A, Bergman K, Carloni L (2010) Phoenixsim: A simulator for physical-layer analysis of chip-scale photonic interconnection networks. In: Design, Automation Test in Europe Conference Exhibition (DATE), 2010, pp 691 –696

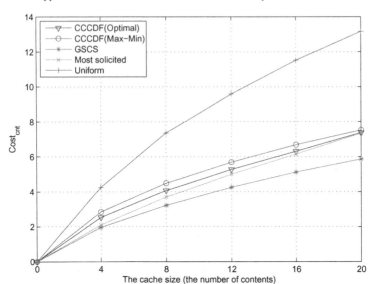

Fig. 2.23 Comparison of overall content caching cost for various caching strategies (Pareto($k = 1.5, x_m = 1$)).

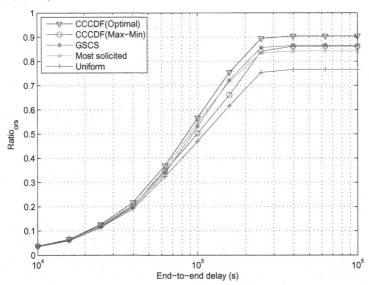

Fig. 2.24 CDFs of end-to-end request serving delay (cache size = 12) (Pareto($k = 1.5, x_m = 1$)).

[11] Costa P, Musolesi M, Mascolo C, Picco GP (2006) Adaptive content-based routing for delay-tolerant mobile ad hoc networks. Technical Report RN-06-08, Department of Computer Science, University College London, London, UK

[12] Costa P, Mascolo C, Musolesi M, Picco G (2008) Socially-aware routing for publish-subscribe in delay-tolerant mobile ad hoc networks. Selected Areas in Communications, IEEE Journal on 26(5):748 –760, DOI 10.1109/JSAC.2008.080602

[13] Dietrich I, Dressler F (2009) On the lifetime of wireless sensor networks. ACM Trans Sen Netw 5(1):1–39, DOI http://doi.acm.org/10.1145/1464420.1464425

[14] Groenevelt R, Nain P, Koole G (2005) The message delay in mobile ad hoc networks. Performance Evaluation 62(1-4):210 – 228, DOI DOI: 10.1016/j.peva.2005.07.018, performance 2005

[15] Guidec F, Maheo Y (2007) Opportunistic content-based dissemination in disconnected mobile ad hoc networks. In: Mobile Ubiquitous Computing, Systems, Services and Technologies, 2007. UBICOMM '07. International Conference on, Papeete, French Polynesia (Tahiti), pp 49–54, DOI 10.1109/UBICOMM.2007.23

[16] Haas Z, Small T (2006) A new networking model for biological applications of ad hoc sensor networks. Networking, IEEE/ACM Transactions on 14(1):27–40, DOI 10.1109/TNET.2005.863461

[17] Haillot J, Guidec F (2008) A protocol for content-based communication in disconnected mobile ad hoc networks. In: Advanced Information Networking and Applications, 2008. AINA 2008. 22nd International Conference on, GinoWan, Okinawa, Japan, pp 188–195, DOI 10.1109/AINA.2008.82

[18] Huang X, Zhai H, Fang Y (2008) Robust cooperative routing protocol in mobile wireless sensor networks. Wireless Communications, IEEE Transactions on 7(12):5278–5285, DOI 10.1109/T-WC.2008.060680

[19] Huang XL, Bensaou B (2001) On max-min fairness and scheduling in wireless ad-hoc networks: analytical framework and implementation. In: MobiHoc '01: Proceedings of the 2nd ACM international symposium on Mobile ad hoc networking & computing, ACM, Long Beach, CA, USA, pp 221–231

[20] Hui P, Leguay J, Crowcroft J, Scott J, Friedmani T, Conan V (2006) Osmosis in pocket switched networks. In: Communications and Networking in China, 2006. ChinaCom '06. First International Conference on, Beijing, China, pp 1–6, DOI 10.1109/CHINACOM.2006.344671

[21] Hunter T, Nosratinia A (2006) Diversity through coded cooperation. Wireless Communications, IEEE Transactions on 5(2):283–289, DOI 10.1109/TWC.2006.1611050

[22] Ibrahim A, Han Z, Liu K (2008) Distributed energy-efficient cooperative routing in wireless networks. Wireless Communications, IEEE Transactions on 7(10):3930–3941, DOI 10.1109/T-WC.2008.070502

[23] Janani M, Hedayat A, Hunter T, Nosratinia A (2004) Coded cooperation in wireless communications: space-time transmission and iterative decoding. Signal Processing, IEEE Transactions on 52(2):362–371, DOI 10.1109/TSP.2003.821100

[24] Khandani A, Abounadi J, Modiano E, Zheng L (2007) Cooperative routing in static wireless networks. Communications, IEEE Transactions on 55(11):2185–2192, DOI 10.1109/TCOMM.2007.908538

[25] Krifa A, Baraka C, Spyropoulos T (2008) Optimal buffer management policies for delay tolerant networks. In: Sensor, Mesh and Ad Hoc Communications and Networks, 2008. SECON '08. 5th Annual IEEE Communications Society Conference on, San Francisco, CA, USA, pp 260–268, DOI 10.1109/SAHCN.2008.40

[26] Krifa A, Barakat C, Spyropoulos T (2008) An optimal joint scheduling and drop policy for delay tolerant networks. In: World of Wireless, Mobile and Multimedia Networks, 2008. WoWMoM 2008. 2008 International Symposium on a, Newport Beach, CA, USA, pp 1–6, DOI 10.1109/WOWMOM.2008.4594889

[27] Le L, Hossain E (2008) Cross-layer optimization frameworks for multihop wireless networks using cooperative diversity. Wireless Communications, IEEE Transactions on 7(7):2592–2602, DOI 10.1109/TWC.2008.060962

[28] Leguay J, Lindgren A, Scott J, Friedman T, Crowcroft J (2006) Opportunistic content distribution in an urban setting. In: CHANTS '06: Proceedings of the 2006 SIGCOMM workshop on Challenged networks, ACM, Pisa, Italy, pp 205–212, DOI http://doi.acm.org/10.1145/1162654.1162657

[29] Lenders V, Karlsson G, May M (2007) Wireless ad hoc podcasting. In: Sensor, Mesh and Ad Hoc Communications and Networks, 2007. SECON '07. 4th Annual IEEE Communications Society Conference on, San Diego, California, USA, pp 273–283, DOI 10.1109/SAHCN.2007.4292839

[30] Lindgren A, Doria A, Schelén O (2004) Probabilistic routing in intermittently connected networks. In: Service Assurance with Partial and Intermittent Resources, Lecture Notes in Computer Science, vol 3126/2004, Springer Berlin / Heidelberg, pp 239–254, DOI 10.1007/b99076

[31] Liu H, Zhang B, Mouftah H, Shen X, Ma J (2009) Opportunistic routing for wireless ad hoc and sensor networks: Present and future directions. Communications Magazine, IEEE 47(12):103–109, DOI 10.1109/MCOM.2009.5350376

[32] Liu P, Tao Z, Lin Z, Erkip E, Panwar S (2006) Cooperative wireless communications: a cross-layer approach. Wireless Communications, IEEE 13(4):84–92, DOI 10.1109/MWC.2006.1678169

[33] Ma Y, Jamalipour A (2009) Cooperative content dissemination in intermittently connected networks. In: Communications, 2009. ICC '09. IEEE International Conference on, Dresden, Germany, pp 1–5, DOI 10.1109/ICC.2009.5198860

[34] Ma Y, Jamalipour A (2010) A cooperative cache-based content delivery framework for intermittently connected mobile ad hoc networks. Wireless Communications, IEEE Transactions on 9(1):366–373, DOI 10.1109/TWC.2010.01.090775

[35] Ma Y, Kibria M, Jamalipour A (2008) Cache-based content delivery in opportunistic mobile ad hoc networks. In: Global Telecommunications Conference, 2008. IEEE GLOBECOM 2008. IEEE, New Orleans, LA, USA, pp 1–5, DOI 10.1109/GLOCOM.2008.ECP.153

[36] Ma Y, Rubaiyat Kibria M, Jamalipour A (2008) Optimized routing framework for intermittently connected mobile ad hoc networks. In: Communications, 2008. ICC '08. IEEE International Conference on, Beijing, China, pp 3171–3175, DOI 10.1109/ICC.2008.597

[37] Macedo M, Grilo A, Nunes M (2009) Distributed latency-energy minimization and interference avoidance in tdma wireless sensor networks. Comput Netw 53(5):569–582, DOI http://dx.doi.org/10.1016/j.comnet.2008.10.015

[38] Marbach P (2003) Priority service and max-min fairness. Networking, IEEE/ACM Transactions on 11(5):733–746, DOI 10.1109/TNET.2003.818196

[39] May M, Lenders V, Karlsson G, Wacha C (2007) Wireless opportunistic podcasting: implementation and design tradeoffs. In: CHANTS '07: Proceedings of the second ACM workshop on Challenged networks, ACM, Montreal, Quebec, Canada, pp 75–82, DOI http://doi.acm.org/10.1145/1287791.1287806

[40] Meyer H, Hummel KA (2009) A geo-location based opportunistic data dissemination approach for manets. In: CHANTS '09: Proceedings of the 4th ACM workshop on Challenged networks, ACM, Beijing, China, pp 1–8, DOI http://doi.acm.org/10.1145/1614222.1614224

[41] Muhl G, Ulbrich A, Herrman K (2004) Disseminating information to mobile clients using publish-subscribe. Internet Computing, IEEE 8(3):46 – 53, DOI 10.1109/MIC.2004.1297273

[42] Ng DKW (1994) Grey system and grey relational model. SIGICE Bull 20(2):2–9, DOI http://doi.acm.org/10.1145/190690.190691

[43] Nosratinia A, Hunter T, Hedayat A (2004) Cooperative communication in wireless networks. Communications Magazine, IEEE 42(10):74–80, DOI 10.1109/MCOM.2004.1341264

[44] Patterson L (2006) The technology underlying podcasts. Computer 39(10):103 –105, DOI 10.1109/MC.2006.361

[45] Payton J, Julien C, Roman GC, Rajamani V (2010) Semantic self-assessment of query results in dynamic environments. ACM Trans Softw Eng Methodol 19(4):1–33, DOI http://doi.acm.org/10.1145/1734229.1734231

[46] Radunovic B, Le Boudec JY (2007) A unified framework for max-min and min-max fairness with applications. Networking, IEEE/ACM Transactions on 15(5):1073–1083, DOI 10.1109/TNET.2007.896231

[47] Saaty TL (2000) Fundamentals of Decision Making and Priority Theory With the Analytic Hierarchy Process, Analytic Hierarchy Process Series, vol 6. RWS Publications

[48] Scott J, Gass R, Crowcroft J, Hui P, Diot C, Chaintreau A (2006) CRAWDAD trace cambridge/haggle/imote/content (v. 2006-09-15). Http://crawdad.cs.dartmouth.edu/cambridge/haggle/imote/content, accessed in December 2009

[49] Spyropoulos T, Psounis K, Raghavendra CS (2006) Performance analysis of mobility-assisted routing. In: MobiHoc '06: Proceedings of the 7th ACM international symposium on Mobile ad hoc networking and computing, ACM, Florence, Italy, pp 49–60, DOI http://doi.acm.org/10.1145/1132905.1132912

[50] Stamos K, Pallis G, Vakali A, Katsaros D, Sidiropoulos A, Manolopoulos Y (2010) Cdnsim: A simulation tool for content distribution networks. ACM Trans Model Comput Simul 20(2):1–40, DOI http://doi.acm.org/10.1145/1734222.1734226

[51] Sung Y, Misra S, Tong L, Ephremides A (2007) Cooperative routing for distributed detection in large sensor networks. Selected Areas in Communications, IEEE Journal on 25(2):471–483, DOI 10.1109/JSAC.2007.070221

[52] Tan K, Zhang Q, Zhu W (2003) Shortest path routing in partially connected ad hoc networks. In: Global Telecommunications Conference, 2003. GLOBECOM '03. IEEE, San Francisco, CA, USA, vol 2, pp 1038–1042 Vol.2, DOI 10.1109/GLOCOM.2003.1258396

[53] Vakali A, Pallis G (2003) Content delivery networks: status and trends. Internet Computing, IEEE 7(6):68 – 74, DOI 10.1109/MIC.2003.1250586

[54] Weingartner E, vom Lehn H, Wehrle K (2009) A performance comparison of recent network simulators. In: Communications, 2009. ICC '09. IEEE International Conference on, Dresden, Germany, pp 1–5, DOI 10.1109/ICC.2009.5198657

[55] Xu Mw, Wu Q, Xie Gl, Zhao Yj (2009) The impact of mobility models on mobile ip multicast research. Int J Ad Hoc Ubiquitous Comput 4(3/4):191–200, DOI http://dx.doi.org/10.1504/IJAHUC.2009.024522

[56] Yan Y, Zhang B, Mouftah H, Ma J (2008) Practical coding-aware mechanism for opportunistic routing in wireless mesh networks. In: Communications, 2008. ICC '08. IEEE International Conference on, pp 2871–2876, DOI 10.1109/ICC.2008.541

[57] Yasar AUH, Mahmud N, Preuveneers D, Luyten K, Coninx K, Berbers Y (2010) Where people and cars meet: social interactions to improve information sharing in large scale vehicular networks. In: SAC '10: Proceedings of the 2010 ACM Symposium on Applied Computing, ACM, Sierre, Switzerland, pp 1188–1194, DOI http://doi.acm.org/10.1145/1774088.1774339

[58] Zeng K, Yang Z, Lou W (2009) Location-aided opportunistic forwarding in multirate and multihop wireless networks. Vehicular Technology, IEEE Transactions on 58(6):3032–3040, DOI 10.1109/TVT.2008.2011637

[59] Zhang J, Zhang Q (2008) Cooperative routing in multi-source multi-destination multi-hop wireless networks. In: INFOCOM 2008. The 27th Conference on Computer Communications. IEEE, Phoenix, AZ, USA, pp 2369–2377, DOI 10.1109/INFOCOM.2008.306

[60] Zhang X, Ansari J, Mähönen P (2009) Traffic aware medium access control protocol for wireless sensor networks. In: MobiWAC '09: Proceedings of the 7th ACM international symposium on Mobility management and wireless access, ACM, Tenerife, Canary Islands, Spain, pp 140–148, DOI http://doi.acm.org/10.1145/1641776.1641802

[61] Zhou Z, Zhou S, Cui JH, Cui S (2008) Energy-efficient cooperative communication based on power control and selective single-relay in wireless sensor networks. Wireless Communications, IEEE Transactions on 7(8):3066–3078, DOI 10.1109/TWC.2008.061097

[62] Zhou Z, Zhou S, Cui S, Cui JH (2008) Energy-efficient cooperative commu-nication in a clustered wireless sensor network. Vehicular Technology, IEEE Transactions on 57(6):3618–3628, DOI 10.1109/TVT.2008.918730

[63] Zorzi M, Rao R (2003) Geographic random forwarding (geraf) for ad hoc and sensor networks: multihop performance. Mobile Computing, IEEE Transac-tions on 2(4):337–348, DOI 10.1109/TMC.2003.1255648

Chapter 3
Opportunistic Content Search in Intermittently Connected Mobile Ad Hoc Networks

3.1 Introduction

Apart from designing forwarding strategies to delivery a content from its publisher to their interested consumers [26][27][21][17][18][10][11], content search or content lookup is another fundamental problem that determines the architecture and performance of opportunistic content distribution networks in the ICMANs context. This is largely because only after discovering the information about its interested contents (e.g., content identifiers, related node addresses and even corresponding meta-data), a mobile user can determine which contents to be downloaded and accordingly activate the downloading process over the network. Note that distributed content search schemes have been widely studied since peer-to-peer (P2P) networks become a promising technology for content dissemination over the Internet [3]. Moreover, due to the common characteristics such as decentralized architecture, self-organization and self-healing features between P2P networks and MANETs, similar content search schemes are also extensively investigated for MANETs based mobile content distribution networks [14][19][15][46][25][8][37]. Unfortunately, to the best of knowledge, one of the key assumptions of underlying routing protocols (i.e., the existence of end-to-end routing path between any two nodes) for all such content search schemes becomes untenable in the ICMANs context, since ICMANs experience intermittent connections due to high node mobility, sporadic node density, short transmission range, and so on [54][36].

With all these in mind, this Chapter first introduces a simple content search scheme for opportunistic content distribution over ICMANs. In the proposed scheme, due to intermittent connectivity, all mobile nodes manage their available contents in a fully decentralized manner. As a result, an epidemic content query forwarding mechanism derived from the well-known epidemic routing technology [49] (a well-known plain flooding based mechanism for ICMANs) is presented to flood all content query messages across ICMANs. It allows other nodes (as many as possible) check whether they contain the matching contents or not, since there do not exist any well-known index servers or powerful super nodes for content lookup purpose. The proposed epidemic content query forwarding mechanism is facilitated by opportunistically employment of mobility nature and storage spaces of the intermediate nodes. However, performance of epidemic content query forwarding may be significantly degraded in most realistic cases with constrained buffer space. Therefore, instead of the blindly flooding across ICMANs, an intelligent epidemic content query forwarding mechanism is accordingly investigated to improve search efficiency under restricted resource conditions. The proposed approach allow the intermediate nodes to use mobility statistics and indexing information from other nodes to choose which content query messages can be stored and carried when the buffer space becomes full.

3.2 Content Search for P2P Content Distribution Networks

In P2P content distribution networks, the participating peers are largely connected via *ad hoc* connections over the Internet to provide distributed content and information sharing for audio, video, data or anything in digital format. By forming an overlay network on top of the Internet, P2P content distribution allows peers to publish, search and download contents from each other. In addition to these functions, some of P2P content delivery networks also provide some other important features such as fairness, security, scalability and performance guarantees.

In terms of "P2P", there are different definitions varying from the strictest definition of pure P2P, where all peers are completely equivalent in terms of functionalities, to a much broader and widely accepted definition that only requires the peers to be able to employ resources from other peers, thereby encompassing other P2P systems, where centralized servers or super peers are utilized. For these different types of P2P content delivery networks, different content management architectures and content search mechanisms are proposed in the literature [8][41][38]. By considering how to organize the participating peers to build overlay network and how to place the contents, these P2P content delivery networks are classified as unstructured and structured P2P networks.

3.2.1 Unstructured P2P

In the unstructured P2P content delivery networks, the placement of available contents is completely unrelated to the overlay topology. However, by considering how to provide efficient content search, the unstructured P2P systems can be further categorized as fully decentralized, centralized and partially centralized.

3.2.1.1 Fully decentralized

Gnutella is a representative example of the fully decentralized and unstructured P2P networks. In Gnutella, all peers perform exactly the same functionalities and there is no central coordination of their activities. Moreover, all peers in Gnutella are connected nondeterministically and all available contents are just stored and indexed locally to handle highly-transient node populations. Accordingly, the simplest content search scheme for Gnutella is to flood content queries across the network in a breadth-first or depth-first manner to locate all desired contents. To limit the spread of queries through the network, a time-to-live (TTL) field is employed by content queries. However, even TTL is employed, the flood-based content search mechanisms are not scalable for the networks with large scale [3].

As can be seen, scalability is the most important issue for the fully decentralized and unstructured P2P networks. Hence, a number of strategies have been proposed to improve the search efficiency and to avoid high network traffic caused by flooding. For example, a query algorithm based on multiple random works is proposed in [31] while another probabilistic scalable P2P search is introduced in [34]. Moreover, instead of flooding, [12] employs routing indices (RIs) to forward content search over the network. Given a query, a RI returns a list of neighbors, ranked according to their goodness for the query, to allow the peer to determine which neighbor to forward the query to. The notion of goodness reflects the number of contents in the "nearby" peers. Similar intelligent search mechanisms (ISMs) [53] can be found in [23][52].

Apart from the original ad hoc connections to other random participating peers, in GES [56], a peer may also have semantic connections to other relevant peers. The relevance of the two peers is determined by their vectors generated from their contained documents, while the semantic connections can eventually organize relevant peers into semantic groups. Accordingly, given a content query, GES first directs the query through random connections by biased walks to the most relevant semantic groups and then floods the query over semantic connections to the most relevant peers. Similarly, in [43], in addition to Gnutella connections, peers that share similar interests are also connected with each other by interest-based shortcuts.

Another approach to reduce query traffic in fully decentralized and unstructured P2P networks is to proactively place several replications of contents or contents related meta-data on some other peers over the networks. For example, uniform, proportional and square-root replications of contents are investigated in [9]. Since increasing the number of replicas of a content decreases the traffic generated by the queries for the file, the fundamental trade-off between the number of replicas and query traffic is investigated in [48].

3.2.1.2 Centralized

In contrast to the fully decentralized and unstructured P2P networks, where the contents are indexed locally, a typical centralized P2P content delivery networks employs a centralized server to manage all users and to index all available contents stored by them. Whenever a new peer wishes to join the network has to register itself and report all contents it has to the server, while the server can subsequently index all these contents along with meta-data descriptions for other peers to query. On receiving a content query from a user, the index server can return a list of matching contents with their exact locations. As a result, the user can directly contact with the peers that hold the matching contents to activate download process. Note that Napster is a representative example of such kind of P2P content delivery networks with a centralized index server, where content search is performed over a centralized directory while content download still occurs in a P2P fashion.

As can be seen, it is easy to implement such centralized P2P networks, while content can also be located quickly and efficiently. Moreover, different search mechanisms varying from exact content match to the more complex content search mechanisms such as keyword-based content search, full-text search and even semantic search can also be implemented easier by the centralized index server. However, the main problem of this centralized architecture is the single point of failure problem for the centralized server, since all the essential information for content search and download is controlled by the single company, institution, or user, who is maintaining the server. In addition, these system may also suffer the scalability issues, since the size of content database and the performance of content query are limited by the capacity of the index server.

3.2.1.3 Partially centralized

The aim of partially centralized P2P architecture is to exploit and to take advantage of the inherent heterogeneity of P2P systems to present a cross between fully decentralized and centralized P2P systems. In the partially centralized P2P systems (also referred to as the hybrid P2P systems or super-peer P2P systems), the concept of super-peer is utilized to introduce hierarchy into the network. Different from a fully decentralized system, where all of the peers will be mostly equally loaded, regardless their capabilities of CPU power, storage space and even bandwidth, a partially centralized P2P system employs some super-nodes with higher capabilities to act as locally centralized index servers to their surrounding peers and proxy content queries on behalf of these peers. Even though most of the peers are only connected to their super-peers, these super-peers are connected to each other as peers in a fully decentralized system.

KaZaA based on FastTrack structure is a typical instance of super-peer P2P networks, while the super-peer based approach is also employed by Gia [8] to make Gnutella-like P2P systems scalable. A similar approach can also be found in [51], while the designing issues of a partially centralized P2P networks are addressed in [5] and a dynamic layer management in super-peer architectures is investigated in [50]. Moreover, a semantic overlay clustering technology is introduced in [30] to facilitate exact, partial, similar and even ontology content search within super-peer P2P networks. In addition, a novel category overlay infrastructure is proposed in [29] to support category overlay search over super-peer P2P networks while Plexus, a scalable P2P protocol is investigated in [2] to enable efficient subset search for P2P networks by employing a partially decentralized architecture.

3.2.2 Structured P2P

To address the scalability issues occurring in the unstructured networks, structured networks try to create network topology and place the contents based on some specific rules so that content queries can be efficiently forwarded to the node with desired contents. Most of such fully decentralized but structured networks employ the distributed hash table (DHT) as the underlying technology for topology construction and content placement.

For example, in PAST (a large-scale peer-to-peer persistent storage utility) system [42], by using DHT, the peers and contents are each assigned uniformly distributed identifiers while replicas of a content are placed at the k peers whose identifiers are most close to the identifier of the content. Note that PAST is layered on top of Pastry [41], an efficient P2P request routing and content location scheme ensuring that the reliable route path length for a content request to its appropriate peers is logarithmic in the total number of peers under normal operation by building the routing table based on identifier prefixes. OceanStore [40] based on Tapestry [55] and cooperative file system (CFS) [13] on top of Chord [44] are another two structured P2P systems providing global, transactional and persistent storage service. In addition, CAN [38], a scalable content addressable network conceptually built over a d-dimensional Cartesian space, is also can be employed to provide hash table-like functionality for P2P content distribution. Even though structured P2P systems can offer a scalable content queries, it is difficult to maintain the system structures when peers are joining and leaving ar a high rate.

Note that in all these DHT-based content storage systems, content only can be retrieved by the users, who already have the single and unique identifier of the corresponding content. In other words, only exact name match is supported while no functionality is provided for keyword-based content search. As a result, [39] introduces a DHT-based distributed search engine to map keyword queries to the peers, which store the unique content identifiers of the relevant contents. Moreover, [22] builds a distributed index scheme on an r-dimensional hypercube vector space to enable keyword/attribute search in DHT-based P2P systems. According to its keywords, a content can be mapped to an r-bit vector, which represents a point in an r-dimensional hypercube space, while by further mapping the hypercube space to the underlying DHT network, this content has one, and only has one peer to index it. As a result, locating an content according to its keywords is simply routing a message to the peer that can handle such keywords. In addition, in [16], contents are represented using attribute-value pairs (AV-pairs) while rendezvous points (RPs) are used for content registration and query resolution. Note that RP set for content name registration is selected by hashing each corresponding AV-pair.

More complex content search has also been studied for structured P2P networks. For example, pSearch [47] provides content-based full-text search over CAN [38] by using vector space model (VSM) to represent contents and queries as term vector in the Cartesian space and utilizing latent semantic indexing (LSI) to generate content semantics.

3.3 P2P Content Search over MANETs

Due to the common characteristics such as decentralized architecture, self-organization and self-healing features between P2P networks and MANETs, P2P content search over MANETs are also extensively investigated for mobile content distribution. For example, [14][19][15] are proposed for P2P content search in MANETs by employing the fully decentralized architecture. In [14], two efficient search schemes employing query filtering/gossiping and adaptive hop-limited search, respectively, to offer scalable performance for P2P content sharing over MANETs. Reference [19] employs Swarm Intelligence to design an efficient and scalable P2P content sharing system for MANETs, while by identifying the regions of the network where the required contents are more likely to be stored. Reference [15] devises an efficient query/response propagation algorithm to efficiently retrieve user contents in MANETs.

In order to employ the partially centralized P2P architectures to facilitate content search over MANETs, [46] integrates FastTrack and AODV routing protocol into a common framework to save content lookup overhead, while RAON [25] incorporates all design features of Gia [8] but performs query forwarding decisions based on link instability and power constraints to improve the success rate and delay of content search in the MANETs context. Moreover, a mobility-aware file discovery control scheme is proposed in [20] for a super-peer architecture over the wireless mobile networks. In addition, index server optimization to reduce content search overhead is studied in [35].

For structured P2P approaches, Ekta [37] integrates Pastry with dynamic source routing (DSR) at the network layer to lookup and maintain routes with reduced overhead, while M-CAN [1] is an extension of CAN. Some other DHT-based search mechanisms are also investigated by [7][45] and comparison of several cross-layer approaches such as Ekta and FastTrack over AODV for P2P systems over MANETs can be found in [6].

3.4 Epidemic Content Search over ICMANs

In the proposed mechanism, apart from the content itself, every content is assumed to contain a unique content identifier and a meta-data file with corresponding keywords. Due to intermittent connectivity, to avoid high maintenance overhead for structured, centralized and partially centralized content search, the fully decentralized architecture is employed at here. In other words, in ICMANs, every mobile node indexes all of its own stored contents (published and downloaded contents) based on their corresponding keywords. As a result, whenever a node wants to download some interested contents without any information about the corresponding content identifiers, it can facilitate a content query based on the related keywords across the network. On receiving such a content query message, each intermediate node performs a keyword-based search through its stored indices, while if any matching content exists, a response message is accordingly generated and forwarded back. The response message includes the content identifiers and the meta-data files of the matching contents. On receiving these information, the node can eventually determine which contents it would like to download and accordingly requests them over the network based on the received content identifiers.

As mentioned earlier, since the contents are indexed locally in the fully decentralized architecture, to allow other nodes (as many as possible) check whether they contain the matching contents or not, the content queries are flooded across the network. Moreover, by considering the intermittently connected manner of ICMAN, the mobility of the nodes are utilized to diffuse content queries over ICMANs. In other words, epidemic content query diffusion process is facilitated to handle the underlying intermittent connectivity. For example, as shown in Fig. 3.1, despite the fact that node A cannot directly encounter nodes D and E, the mobility and the storage spaces of node B, C and D are employed by node A to eventual deliver its content query message to them.

To perform epidemic content search, every issued query message includes node ID, query ID, interested keywords, TTL, timestamp and expiry time. Node ID and query ID together represent each unique content query message, while expiry time and TTL are given to address content searching scalability in the ICMANs context. Note that for a given query message, if the expiry time is passed, it should be removed by all possible intermediate nodes from the storage. In addition, the TTL field is employed to limit the search scope of a request.

Due to the intermittent connectivity, content query forwarding and content search only happen whenever two mobile nodes encounter each other. After connection is established between two encountering nodes, the entire procedure can be summarized as follows:

1. Firstly, each node checks its buffer space and removes all expired content query messages.
2. Then each node sends all query messages (stored or cached) to its encountering node.
3. On receiving a query message, the node executes the keyword-based search through its locale index server.
4. After that, if such a query message is a new search request for the node and its TTL limitation is not exceeded, it will be inserted into the locale cache space for further forwarding.

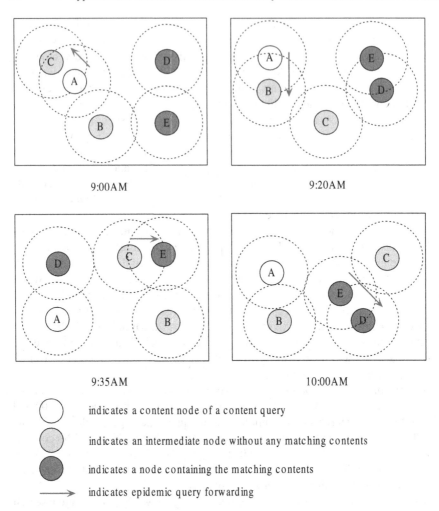

Fig. 3.1 Epidemic P2P content search mechanism in ICMANs.

As mentioned earlier, for epidemic routing, a node may store a message in its buffer and carry it for a long time until another possible forwarding opportunity arises while multiple replicas of a message may be stored and carried by different nodes to increase the serving probability. Such combination of message replications and long time storage involves a very high storage cost, therefore performance of epidemic routing may be significantly degraded in most realistic cases with constrained buffer space. Similar situation can be suffered by epidemic content query forwarding.

To resolve this issue for epidemic routing, various buffer management policies were proposed in the literature. Some of them can be easily employed by the proposed epidemic content search scheme. For instance, First In First Out (FIFO) policy [28], where the message that was first entered the storage space is the first message to be dropped when the buffer is full, and Evict Shortest Lifetime First (SHLI) [28], where the message that has shortest lifetime is the first message to be discarded, can be directly utilized for epidemic content query forwarding. While some others such as Global Knowledge based Drop (GBD) policy [24], where each message is evaluated based on collected information such as mobility characteristics before making any message discarding decisions, cannot be employed directly. This is because each message forwarded by epidemic routing only has one destination and such destination is known by the intermediate nodes, while in epidemic content search scheme, each content query message only has the keywords, which may involve more than one destined nodes without any information about their addresses. However, it does not mean similar idea cannot be implemented for the proposed epidemic content search scheme. In the next section, in order to improve search performance, an intelligent epidemic content search scheme with mobility statistics based cache management policy is described in detail.

3.5 Intelligent Epidemic Content Search over ICMANs

As mentioned earlier, in most realistic cases with the restricted cache spaces, store and carrying opportunity cannot be provided to every content search request from other peers, therefore, one basic issue that has to be addressed for epidemic content search scheme is "Which content query messages should be dropped when the cache is full?" To answer this question, each available request has to be evaluated and based on the evaluation result, an intermediate node can determine which queries to be discarded when the cache space of the node is full. The objective of such evaluation is to try to utilize the constrained cache spaces more efficiently so that during a certain period, the network can serve more content queries while more matching contents can be discovered for each content query. Note that each node is assumed to have an unlimited buffer space for its own content query messages while a restricted cache space is reserved to temporarily store such query messages from any other nodes in the network i.e, cache management policy is only carried out for the cached queries belonging to other nodes.

To facilitate the evaluation process, the essential information that a node needs to know when it wants to make any reasonable evaluation for each content query is discussed firstly. Then, some assumptions regarding the information required by content search service itself and the mobility characteristics of the mobile nodes in the ICMANs context are illustrated. After that, the evaluation of each query message based on such essential information is given in detail. Finally, the section explains how the corresponding information can be collected by every node.

As can be seen in Fig. 3.2, if any node can build up such relationship map among keywords, content identifers and node identifiers, it can subsequently learn how many matching contents exist over the network and where to find them when it receives any content query message. Moreover, the node can further estimate its

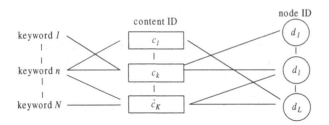

Fig. 3.2 Relationship map among keyword, content ID and node ID

discovering probability for any possible matching content with several known locations, if it also has the related historical encountering statistics. Accordingly, the node can eventually get an expected serving ratio, an expected value indicating what percentage of the matching contents that can be discovered in the future, if it offers its storing opportunity to such a content query message. Thus, to improve the overall searching performance, such expected serving ratio can be exploited by epidemic content search mechanism to determine which query messages should be kept and which query messages should be dropped whenever its cache space becomes full.

To carry out the expected serving ratio for content query evaluation, the same mobility assumption as mentioned in the previous chapters is considered. That is the inter-encountering time $\tau_{i,j}$ between each pair of mobile nodes (i, j) follows the exponential distribution. As a result, the meeting rate $\lambda_{i,j}$ between (i, j) is represented as $1/E[\tau_{i,j}]$. Apart from the mobility assumption, the following assumptions are also made for content query message evaluation. Firstly, there is assumed a content query message m at an instant t, which has been performed by node x while is subsequently waiting for a caching decision from node x. Based on its expiry time e_m, its remaining time (r_m) before it becomes expired is

$$r_m = e_m - t$$

Moreover, according to the maintained relationship map (Fig. 3.2), node x can find that there are total $\bar{K}$ matching contents over the network, while as shown in Fig. 3.3, $K(K \leq \bar{K})$ of them are not held by node x. Note that $\bar{K} - K$ represents the number of the matching contents that are discovered by node x from its local index server. These K matching contents are defined as $\{c_{m,k}\}, k \in [1, K]$ while each $c_{m,k}$ is located in $L_{m,k}$ nodes, the IDs of which are defined as $\{d_{m,k,l}\}, l \in [1, L_{m,k}]$.

Based on the above assumptions, the evaluation function $U(m)$ can be carried out as follows. Firstly, due to the exponential distribution, for a content query m, the probability that node x cannot discover the matching content $c_{m,k}$ from a certain location $d_{m,k,l}$ during the remaining time r_m will be:

$$p_{x,d_{m,k,l}} = \exp(-r_m \lambda_{x,d_{m,k,l}}) \tag{3.1}$$

Accordingly, for query m, the probability that peer x can discover the matching content $c_{m,k}$ from the network before its expiry time e_m can be represented as follows:

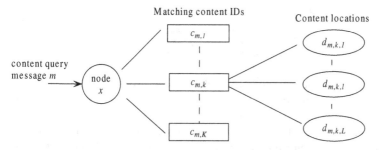

Fig. 3.3 Content search results performed by node x for a content query message m

$$p_{x,c_{m,k}} = 1 - \prod_{l=1}^{L_{m,k}} p_{i,d_{m,k,l}} \tag{3.2}$$

$$= 1 - \prod_{l=1}^{L_{m,k}} \exp(-r_m \lambda_{x,d_{m,k,l}}) \tag{3.3}$$

$$= 1 - \exp(-r_m \sum_{l=1}^{L_{m,k}} \lambda_{x,d_{m,k,l}}) \tag{3.4}$$

Moreover, if a random variable $X_{m,k}$ is defined as the success of discovering content $c_{m,k}$ by node x during the remaining time r_m i.e.,

$$X_{m,k} = \begin{cases} 1, & \text{node } x \text{ can discover } c_{m,k} \text{ for } m \text{ during } r_m; \\ 0, & \text{otherwise} \end{cases}$$

the expected value of $X_{m,k}$ is represented as

$$E(X_{m,k}) = Pr(X_{m,k} = 1) = p_{x,c_{m,k}} \tag{3.5}$$

This is because that $X_{m,k}$ follows a Bernoulli distribution. Therefore, for such K matching independent contents, if a random variable X_m is defined as total contents that can be discovered successfully by node x for request m during the remaining time r_m, the expected value of X_m can be derived as follows:

$$E[X_m] = E[\sum_{k=1}^{K} X_{m,k}] = \sum_{k=1}^{K} E[X_{m,k}]$$

$$= K - \sum_{k=1}^{K} \exp(-r_m \sum_{l=1}^{L_{m,k}} \lambda_{x,d_{m,k,l}}) \tag{3.6}$$

Finally, the evaluation function $U(m)$, which can be employed to get the expected discovering ratio offered by peer x for content query message m during the remaining time r_m, will be:

$$U(m) = E[X_m]/\bar{K}$$

$$= \frac{K}{\bar{K}} - \frac{1}{\bar{K}} \sum_{k=1}^{K} \exp(-r_m \sum_{l=1}^{L_{m,k}} \lambda_{x,d_{m,k,l}}) \tag{3.7}$$

Based on (3.7), an intermediate node can calculate the expected discovering ratio for each possible content query message and accordingly drop the candidates with lower values when the cache space is full.

As can be seen from (3.7), to perform $U(m)$ for request evaluation, each node needs to collect the inter-encountering time for any other node it encountered in the history and accordingly calculates the corresponding meeting rates. In addition, every node has to construct the relationship map illustrated in Fig. 3.2. There are different ways for this. For example, a node can construct it by sniffing any received query response messages, which include required content identifers and related keywords. Different from such passive way, alternatively, each node can broadcast its stored indices (only content identifers and related keywords) with an expiry time over the network periodically to allow others to learn it. To reduce the related overhead, such broadcast can be limited according to the pre-defined depth and breadth restrictions. In other words, each indices update message includes node ID, timestamp, expiry time and TTL. In the intelligent epidemic content search mechanism, the proactive approach is considered. Furthermore, if a node cannot receive any updated indices after the expiry time, such indices will be removed by the node. Note that based on such relationship map, even more complicated boolean expression based content query still can be evaluated by any intermediate node.

Since the query message evaluation is involved in the intelligent epidemic content search, the previously introduced procedure for any two encountering node is resummarized as below.

1. Both nodes update their λ for all past encounters firstly.
2. Then, both nodes generate their index update messages with timestamp and such latest update messages as well as all other cached unexpired update messages are forwarded to each other based on their TTL limitations. On receiving such index update messages, each node only caches the latest update messages for every other node based on the timestamp and reconstruct the relationship map.
3. After that, both nodes check their buffer space and removes all expired content query messages.
4. Subsequently, each node sends all query messages (stored or cached) to its encountering node.
5. On receiving these content query messages, the node executes them through its locale index server.
6. Finally, all newly received and already cached content query messages are evaluated based on (3.7) and the cache spaces are only allocated to the query messages with higher expected discovering ratios. The query messages without any caching opportunities are discarded by the nodes.

3.6 Performance Evaluation

3.6.1 Simulation Methodology

Performance of the proposed epidemic content search mechanisms (basic and intelligent) across various cache conditions is evaluated by using OMNeT++ discrete event simulator while different independent seeds are utilized to provide average results. As mentioned earlier, cache has restricted space reserved to temporarily store content search requests from any other peers. During the simulation, total of 64 mobile peers are considered while an exponential distribution based synthetic mobility model is employed. Which pair of nodes has an opportunity to encounter each other during the whole simulation is randomly determined and such probability is set as 0.4. Once two nodes can encounter each other, its inter-encountering time (i.e., $E[\tau_{i,j}]$) follows a Gaussian distribution with $N(1500\,sec, 300\,sec)$.

Simulation considers a total of 512 unique contents with 128 related keywords. Note that the simulation just focuses on how content query messages can be diffused over ICMANs by epidemic content search schemes and how the evaluation process can affect the overall searching performance (rather than the keyword-based content search itself). Therefore, to simplify the simulation, each content only selects one related keyword randomly. Moreover, in most realistic cases, people collect and store contents based on their interests and a certain interest may just attract a group of users. Therefore, to simplify the simulation, the location of each keyword is determined instead of determining the location of each content directly. The simulation first chooses the number of locations for each keyword, which follows a Gaussian distribution with $N(2, 1)$ in the simulation i.e., each keyword related contents is collected by a small group of users. Then, each exact keyword location is selected from all nodes randomly. After that, the owners of every corresponding content are only chosen from such keyword related locations. In addition, for each content, more than one locations can be selected based on the aforementioned rules.

The whole simulation continues for 4 hours i.e., 14400 sec where after the first 1 hour, the system generates a content search request every 3 sec for another 9600 sec. To simplify the simulation, each content query message only considers one keyword, which is randomly selected from the previous 128 keywords. Such an assumption will not affect the simulation results, since as mentioned earlier, the proposed epidemic content search schemes naturally supports the boolean expression based keyword search. Moreover, the source peer ID is randomly chosen from the nodes that do not contain any matching contents. In addition, the expiry time of each content query is given as 1200 sec, while the TTL limitation is given as 7 hops. Note that in the simulation every peer periodically broadcasts its keyword indices over the network to facilitate the intelligent epidemic content search mechanism. For such indices broadcast, the depth and breadth limitations are given as 3 and 4 peers, respectively.

3.6.2 Performance Measurements

Note that the corresponding response messages already have their well-known destinations, and accordingly, they can be delivered in the network by using any existing SCF routing technologies (e.g., [49][24][4][33][32]). Therefore, for a content query message, it is assumed to be successfully served when one of the matching content holders receives it rather than the source node receives the response messages. Accordingly, content query served ratio ($Ratio_{srvd}$), a main metric proposed to evaluate different schemes, can be defined as follows:

$$Ratio_{srvd} = \frac{num_{srvd}}{num_{total}} \qquad (3.8)$$

where num_{srvd} and num_{total} indicate "total successfully served content queries" and "total generated content queries", respectively. Moreover, to investigate how successfully a content query is served (i.e., how many unique matching contents are discovered before its expiry time), the matching contents average discovered ratio ($Ratio_{avr-dscvrd}$) is defined as follows:

$$Ratio_{avr-dscvrd} = \frac{1}{num_{srvd}} \sum_{i=1}^{num_{srvd}} \frac{n_i}{N_i} \qquad (3.9)$$

where n_i and N_i indicate "Total discovered contents" and "Total existing matching contents over the network" for the i_{th} served content query, respectively.

3.6.3 Simulation Results

The intelligent epidemic content search scheme is compared with the normal epidemic scheme with FIFO policy and SHLI policy in the simulation. Note that FIFO policy handles the cache space in a FIFO order while SHLI policy remove the requests with shortest lifetime first.

Fig. 3.4 provides a comparison between them in terms of $Ratio_{srvd}$. In addition, the zero cache size implies that any peer does not store and forward the content search requests for all other peers (i.e., the peer only stores its own content search requests and such requests only can be executed by its encountering nodes). Even though, "No caching" is not a real epidemic content search scheme, it can be a benchmark for others. As expected, the performance of these policies improves when the peers employ larger cache to carry requests from others. Moreover, over a range of cache sizes, the intelligent epidemic content search performs better than the basic scheme with FIFO policy and SHLI policy. For example, as can be seen from Fig. 3.4, for cache size 12, the served ratio of the intelligent epidemic content search is more than 15% higher than the basic scheme with FIFO and SHLI policies, while comparing to "No caching", the intelligent epidemic search offers even 60% better performance. In other words, comparing with FIFO and SHLI policies, the in-

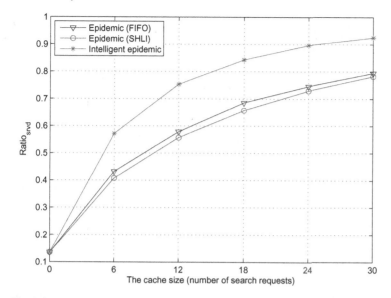

Fig. 3.4 Comparison of requests served ratio for various epidemic content search schemes under different cache conditions.

telligent epidemic search always can use less cache size to offer the same $Ratio_{srvd}$. This is because in such keyword and location correlated scenario, according to (3.7), the characteristics of the mobility will affect the evaluation result of each possible content query message, while the considerable differences among such results can be easily employed by the intelligent epidemic search to make discarding decisions.

While Fig. 3.4 considers requests successfully served ratio, Fig. 3.5 investigates what percentage of the matching contents that can be discovered for each successfully served content query. As can be seen, in terms of $Ratio_{avr-dscvrd}$, the intelligent epidemic search also offers better performance than all others. In other words, besides offering higher successfully served ratio, for each successfully served content query message, the intelligent epidemic search also can help nodes find more matching contents than basic epidemic scheme with FIFO and SHLI dropping policies.

Conversely, Fig. 3.6 illustrates cumulative distribution functions (CDFs) of content discovery delay for all search schemes when cache size is 12. Fig. 3.6 shows that even though the intelligent epidemic search can discover more matching contents before the expected deadlines, the performance of its actual end-to-end serving delay is also close to SHLI policy while better than FIFO policy.

3.7 Summary and Conclusions

In order to deploy a content search service for mobile content consumers by utilizing low cost wireless connectivity, this Chapter proposed an epidemic content search

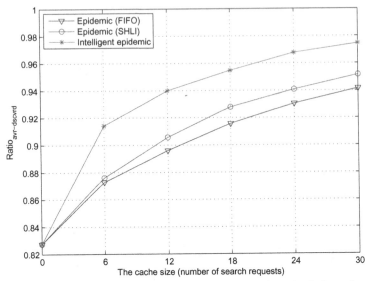

Fig. 3.5 Comparison of the matching contents average discovered ratio for various epidemic content search schemes under different cache conditions.

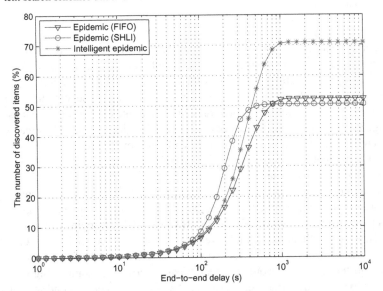

Fig. 3.6 CDFs of the content discovery delay (cache size = 12).

mechanism in the ICMANs context. By considering the nature of intermittent connectivity, the fully decentralized architecture was employed to index the available contents locally and a basic epidemic content search mechanism was accordingly introduced. Moreover, due to the restricted storage spaces in mobile devices, an intelligent epidemic content search mechanism with evaluation-based cache management policy was proposed to improve the buffer usage efficiency by considering the relevant information such as mobility statistics.

References

[1] Ahmed D, Shirmohammadi S (2007) Design issues of peer-to-peer systems for wireless ad hoc networks. In: Networking, 2007. ICN '07. Sixth International Conference on, Sainte-Luce, Martinique, France, pp 26–26, DOI 10.1109/ICN.2007.36

[2] Ahmed R, Boutaba R (2009) Plexus: A scalable peer-to-peer protocol enabling efficient subset search. Networking, IEEE/ACM Transactions on 17(1):130–143, DOI 10.1109/TNET.2008.2001466

[3] Androutsellis-Theotokis S, Spinellis D (2004) A survey of peer-to-peer content distribution technologies. ACM Comput Surv 36(4):335–371, DOI http://doi.acm.org/10.1145/1041680.1041681

[4] Balasubramanian A, Levine B, Venkataramani A (2007) Dtn routing as a resource allocation problem. In: SIGCOMM '07: Proceedings of the 2007 conference on Applications, technologies, architectures, and protocols for computer communications, ACM, Kyoto, Japan, pp 373–384, DOI http://doi.acm.org/10.1145/1282380.1282422

[5] Beverly Yang B, Garcia-Molina H (2003) Designing a super-peer network. In: Data Engineering, 2003. Proceedings. 19th International Conference on, Bangalore, India, pp 49–60

[6] Bisignano M, Di Modica G, Tomarchio O, Vita L (2007) P2p over manet: a comparison of cross-layer approaches. In: Database and Expert Systems Applications, 2007. DEXA '07. 18th International Conference on, Regensburg, Germany, pp 814–818, DOI 10.1109/DEXA.2007.88

[7] Caesar M, Castro M, Nightingale EB, O'Shea G, Rowstron A (2006) Virtual ring routing: network routing inspired by dhts. In: SIGCOMM '06: Proceedings of the 2006 conference on Applications, technologies, architectures, and protocols for computer communications, ACM, Pisa, Italy, pp 351–362, DOI http://doi.acm.org/10.1145/1159913.1159954

[8] Chawathe Y, Ratnasamy S, Breslau L, Lanham N, Shenker S (2003) Making gnutella-like p2p systems scalable. In: SIGCOMM '03: Proceedings of the 2003 conference on Applications, technologies, architectures, and protocols for computer communications, ACM, Karlsruhe, Germany, pp 407–418, DOI http://doi.acm.org/10.1145/863955.864000

[9] Cohen E, Shenker S (2002) Replication strategies in unstructured peer-to-peer networks. In: SIGCOMM '02: Proceedings of the 2002 conference on Applications, technologies, architectures, and protocols for computer communications, ACM, Pittsburgh, Pennsylvania, USA, pp 177–190, DOI http://doi.acm.org/10.1145/633025.633043

[10] Costa P, Musolesi M, Mascolo C, Picco GP (2006) Adaptive content-based routing for delay-tolerant mobile ad hoc networks. Technical Report RN-06-08, Department of Computer Science, University College London, London, UK

[11] Costa P, Mascolo C, Musolesi M, Picco G (2008) Socially-aware routing for publish-subscribe in delay-tolerant mobile ad hoc networks. Selected Areas in Communications, IEEE Journal on 26(5):748 –760, DOI 10.1109/JSAC.2008.080602

[12] Crespo A, Garcia-Molina H (2002) Routing indices for peer-to-peer systems. In: Distributed Computing Systems, 2002. Proceedings. 22nd International Conference on, Vienna, Austria, pp 23–32, DOI 10.1109/ICDCS.2002.1022239

[13] Dabek F, Kaashoek MF, Karger D, Morris R, Stoica I (2001) Wide-area cooperative storage with cfs. In: SOSP '01: Proceedings of the eighteenth ACM symposium on Operating systems principles, ACM, Banff, Alberta, Canada, pp 202–215, DOI http://doi.acm.org/10.1145/502034.502054

[14] Duran A, Shen CC (2004) Mobile ad hoc p2p file sharing. In: Wireless Communications and Networking Conference, 2004. WCNC. 2004 IEEE, Atlanta, GA, USA, vol 1, pp 114–119 Vol.1

[15] Fiore M, Casetti C, Chiasserini CF (2007) Efficient retrieval of user contents in manets. In: INFOCOM 2007. 26th IEEE International Conference on Computer Communications. IEEE, Anchorage, Alaska, USA, pp 10–18, DOI 10.1109/INFCOM.2007.10

[16] Gao J, Steenkiste P (2004) Design and evaluation of a distributed scalable content discovery system. Selected Areas in Communications, IEEE Journal on 22(1):54–66, DOI 10.1109/JSAC.2003.818794

[17] Guidec F, Maheo Y (2007) Opportunistic content-based dissemination in disconnected mobile ad hoc networks. In: Mobile Ubiquitous Computing, Systems, Services and Technologies, 2007. UBICOMM '07. International Conference on, Papeete, French Polynesia (Tahiti), pp 49–54, DOI 10.1109/UBICOMM.2007.23

[18] Haillot J, Guidec F (2008) A protocol for content-based communication in disconnected mobile ad hoc networks. In: Advanced Information Networking and Applications, 2008. AINA 2008. 22nd International Conference on, GinoWan, Okinawa, Japan, pp 188–195, DOI 10.1109/AINA.2008.82

[19] Hoh CC, Hwang RH (2007) P2p file sharing system over manet based on swarm intelligence: A cross-layer design. In: Wireless Communications and Networking Conference, 2007.WCNC 2007. IEEE, Hong Kong, China, pp 2674–2679, DOI 10.1109/WCNC.2007.497

[20] Huang CM, Hsu TH, Hsu MF (2007) Network-aware p2p file sharing over the wireless mobile networks. Selected Areas in Communications, IEEE Journal on 25(1):204–210, DOI 10.1109/JSAC.2007.070120

[21] Hui P, Leguay J, Crowcroft J, Scott J, Friedmani T, Conan V (2006) Osmosis in pocket switched networks. In: Communications and Networking in China,

2006. ChinaCom '06. First International Conference on, Beijing, China, pp 1–6, DOI 10.1109/CHINACOM.2006.344671

[22] Joung YJ, Yang LW, Fang CT (2007) Keyword search in dht-based peer-to-peer networks. Selected Areas in Communications, IEEE Journal on 25(1):46–61, DOI 10.1109/JSAC.2007.070106

[23] Kalogeraki V, Gunopulos D, Zeinalipour-Yazti D (2002) A local search mechanism for peer-to-peer networks. In: CIKM '02: Proceedings of the eleventh international conference on Information and knowledge management, ACM, McLean, Virginia, USA, pp 300–307, DOI http://doi.acm.org/10.1145/584792.584842

[24] Krifa A, Baraka C, Spyropoulos T (2008) Optimal buffer management policies for delay tolerant networks. In: Sensor, Mesh and Ad Hoc Communications and Networks, 2008. SECON '08. 5th Annual IEEE Communications Society Conference on, San Francisco, CA, USA, pp 260–268, DOI 10.1109/SAHCN.2008.40

[25] Lau G, Jaseemuddin M, Ravindran G (2005) Raon: a p2p network for manet. In: Wireless and Optical Communications Networks, 2005. WOCN 2005. Second IFIP International Conference on, Dubai, UAE, pp 316–322, DOI 10.1109/WOCN.2005.1436041

[26] Leguay J, Lindgren A, Scott J, Friedman T, Crowcroft J (2006) Opportunistic content distribution in an urban setting. In: CHANTS '06: Proceedings of the 2006 SIGCOMM workshop on Challenged networks, ACM, Pisa, Italy, pp 205–212, DOI http://doi.acm.org/10.1145/1162654.1162657

[27] Lenders V, Karlsson G, May M (2007) Wireless ad hoc podcasting. In: Sensor, Mesh and Ad Hoc Communications and Networks, 2007. SECON '07. 4th Annual IEEE Communications Society Conference on, San Diego, California, USA, pp 273–283, DOI 10.1109/SAHCN.2007.4292839

[28] Lindgren A, Phanse K (2006) Evaluation of queueing policies and forwarding strategies for routing in intermittently connected networks. In: Communication System Software and Middleware, 2006. Comsware 2006. First International Conference on, Delhi, India, pp 1–10, DOI 10.1109/COMSWA.2006.1665196

[29] Liu X, Wang J, Vuong S (2005) A category overlay infrastructure for peer-to-peer content search. In: Parallel and Distributed Processing Symposium, 2005. Proceedings. 19th IEEE International, Denver, Colorado, USA, pp 204a–204a, DOI 10.1109/IPDPS.2005.3

[30] Löser A, Naumann F, Siberski W, Nejdl W, Thaden U (2004) Semantic overlay clusters within super-peer networks. In: Databases, Information Systems, and Peer-to-Peer Computing, Lecture Notes in Computer Science, vol 2944/2004, Springer Berlin / Heidelberg, pp 33–47, DOI 10.1007/b95270

[31] Lv Q, Cao P, Cohen E, Li K, Shenker S (2002) Search and replication in unstructured peer-to-peer networks. In: ICS '02: Proceedings of the 16th international conference on Supercomputing, ACM, New York, NY, USA, pp 84–95, DOI http://doi.acm.org/10.1145/514191.514206

[32] Ma Y, Jamalipour A (2009) Optimized message delivery framework using fuzzy logic for intermittently connected mobile ad hoc networks.

Wireless Communications and Mobile Computing 9(4):501–512, DOI 10.1002/wcm.693

[33] Ma Y, Rubaiyat Kibria M, Jamalipour A (2008) Optimized routing framework for intermittently connected mobile ad hoc networks. In: Communications, 2008. ICC '08. IEEE International Conference on, Beijing, China, pp 3171–3175, DOI 10.1109/ICC.2008.597

[34] Menasce D (2003) Scalable p2p search. Internet Computing, IEEE 7(2):83–87, DOI 10.1109/MIC.2003.1189193

[35] Ohta C, Ge Z, Guo Y, Kurose J (2004) Index-server optimization for p2p file sharing in mobile ad hoc networks. In: Global Telecommunications Conference, 2004. GLOBECOM '04. IEEE, Dallas, Texas, USA, vol 2, pp 960–966 Vol.2, DOI 10.1109/GLOCOM.2004.1378102

[36] Pelusi L, Passarella A, Conti M (2006) Opportunistic networking: data forwarding in disconnected mobile ad hoc networks. Communications Magazine, IEEE 44(11):134–141, DOI 10.1109/MCOM.2006.248176

[37] Pucha H, Das S, Hu Y (2004) Ekta: an efficient dht substrate for distributed applications in mobile ad hoc networks. In: Mobile Computing Systems and Applications, 2004. WMCSA 2004. Sixth IEEE Workshop on, Lake District National Park, UK, pp 163–173, DOI 10.1109/MCSA.2004.11

[38] Ratnasamy S, Francis P, Handley M, Karp R, Schenker S (2001) A scalable content-addressable network. In: SIGCOMM '01: Proceedings of the 2001 conference on Applications, technologies, architectures, and protocols for computer communications, ACM, San Diego, California, United States, pp 161–172, DOI http://doi.acm.org/10.1145/383059.383072

[39] Reynolds P, Vahdat A (2003) Efficient peer-to-peer keyword searching. In: Middleware '03: Proceedings of the ACM/IFIP/USENIX 2003 International Conference on Middleware, Springer-Verlag New York, Inc., Rio de Janeiro, Brazil, pp 21–40

[40] Rhea S, Eaton P, Geels D, Weatherspoon H, Zhao B, Kubiatowicz J (2003) Pond: The oceanstore prototype. In: FAST '03: Proceedings of the 2nd USENIX Conference on File and Storage Technologies, USENIX Association, San Francisco, CA, pp 1–14

[41] Rowstron A, Druschel P (2001) Pastry: Scalable, decentralized object location, and routing for large-scale peer-to-peer systems. In: Middleware 2001, Lecture Notes in Computer Science, vol 2218/2001, Springer Berlin / Heidelberg, pp 329–350

[42] Rowstron A, Druschel P (2001) Storage management and caching in past, a large-scale, persistent peer-to-peer storage utility. In: SOSP '01: Proceedings of the eighteenth ACM symposium on Operating systems principles, ACM, Banff, Alberta, Canada, pp 188–201, DOI http://doi.acm.org/10.1145/502034.502053

[43] Sripanidkulchai K, Maggs B, Zhang H (2003) Efficient content location using interest-based locality in peer-to-peer systems. In: INFOCOM 2003. Twenty-Second Annual Joint Conference of the IEEE Computer and Communications. IEEE Societies, San Francisco, CA, USA, vol 3, pp 2166–2176 vol.3

[44] Stoica I, Morris R, Karger D, Kaashoek MF, Balakrishnan H (2001) Chord: A scalable peer-to-peer lookup service for internet applications. In: SIGCOMM '01: Proceedings of the 2001 conference on Applications, technologies, architectures, and protocols for computer communications, ACM, San Diego, California, United States, pp 149–160, DOI http://doi.acm.org/10.1145/383059.383071

[45] Takeshita K, Sasabe M, Nakano H (2008) Mobile p2p networks for highly dynamic environments. In: Pervasive Computing and Communications, 2008. PerCom 2008. Sixth Annual IEEE International Conference on, Hong Kong, China, pp 453–457, DOI 10.1109/PERCOM.2008.38

[46] Tang B, Zhou Z, Kashyap A, cker Chiueh T (2005) An integrated approach for p2p file sharing on multi-hop wireless networks. In: Wireless And Mobile Computing, Networking And Communications, 2005. (WiMob'2005), IEEE International Conference on, Montreal, Canada, vol 3, pp 268–274 Vol. 3, DOI 10.1109/WIMOB.2005.1512913

[47] Tang C, Xu Z, Dwarkadas S (2003) Peer-to-peer information retrieval using self-organizing semantic overlay networks. In: SIGCOMM '03: Proceedings of the 2003 conference on Applications, technologies, architectures, and protocols for computer communications, ACM, Karlsruhe, Germany, pp 175–186, DOI http://doi.acm.org/10.1145/863955.863976

[48] Terpstra WW, Kangasharju J, Leng C, Buchmann AP (2007) Bubblestorm: resilient, probabilistic, and exhaustive peer-to-peer search. In: SIGCOMM '07: Proceedings of the 2007 conference on Applications, technologies, architectures, and protocols for computer communications, ACM, Kyoto, Japan, pp 49–60, DOI http://doi.acm.org/10.1145/1282380.1282387

[49] Vahdat A, Becker D (2000) Epidemic routing for partially-connected ad hoc networks. Duke Technical Report CS-2000-06, The Department of Computer Science, Duke University, Durham, NC

[50] Xiao L, Zhuang Z, Liu Y (2005) Dynamic layer management in super-peer architectures. Parallel and Distributed Systems, IEEE Transactions on 16(11):1078–1091, DOI 10.1109/TPDS.2005.137

[51] Xu Z, Hu Y (2003) Sbarc: A supernode based peer-to-peer file sharing system. In: Computers and Communication, 2003. (ISCC 2003). Proceedings. Eighth IEEE International Symposium on, Antalya, Turkey, pp 1053–1058 vol.2, DOI 10.1109/ISCC.2003.1214254

[52] Yang B, Garcia-Molina H (2002) Improving search in peer-to-peer networks. In: Distributed Computing Systems, 2002. Proceedings. 22nd International Conference on, Vienna, Austria, pp 5–14, DOI 10.1109/ICDCS.2002.1022237

[53] Zeinalipour-Yazti D, Kalogeraki V, Gunopulos D (2004) Information retrieval techniques for peer-to-peer networks. Computing in Science & Engineering 06(4):20–26, DOI 10.1109/MCSE.2004.12

[54] Zhang Z (2006) Routing in intermittently connected mobile ad hoc networks and delay tolerant networks: overview and challenges. Communications Surveys & Tutorials, IEEE 8(1):24–37, DOI 10.1109/COMST.2006.323440

[55] Zhao B, Huang L, Stribling J, Rhea S, Joseph A, Kubiatowicz J (2004) Tapestry: a resilient global-scale overlay for service deployment. Selected Areas in Communications, IEEE Journal on 22(1):41–53, DOI 10.1109/JSAC.2003.818784

[56] Zhu Y, Yang X, Hu Y (2005) Making search efficient on gnutella-like p2p systems. In: Parallel and Distributed Processing Symposium, 2005. Proceedings. 19th IEEE International, Denver, Colorado, USA, pp 56a–56a, DOI 10.1109/IPDPS.2005.273

Index